This Old House Bathrooms

A Guide to Design and Renovation

This Old House Bathrooms
A Guide to Design and Renovation

Steve Thomas and Philip Langdon

Little, Brown and Company
Boston   Toronto   London

Also by the authors:

This Old House Kitchens: A Guide to Design and Renovation

First Edition

Library of Congress Cataloging-in-Publication Data

Thomas, Steve.
    This old house bathrooms : a guide to design and renovation / Steve
 Thomas and Philip Langdon.
     p. cm.
    ISBN 0-316-84109-9. — ISBN 0-316-84110-2 (pbk.)
    1. Bathrooms. 2. Bathrooms — Remodeling. I. Langdon,
 Phillip. II. Title.
NK2117.B33T46   1993
643'.52 — dc20   91-25218

10 9 8 7 6 5 4 3 2 1

Drawings: John Murphy
Design: Chris Pullman/WGBH
Art Direction: Pamela Hartford

RRD-OH

Published simultaneously in Canada by Little, Brown & Company
(Canada) Limited

Printed in the United States of America

In Memory of
Will Humphreys

Introduction

Steve Thomas.

Every few minutes, the squalls bore down the mountain, driving the slanting lines of snowflakes before them like minuscule sheep. Some of the snow dusted the ground around us, some stuck on the small, twisted piñon trees and cedar fence, and the rest, at least from our vantage point, was swallowed up in the steaming waters of the pool.

"Incredible," Richard Trethewey marveled. "It's snowing and we're out here swimming. This ain't the way we do it in Boston!" Just then the squall blew past and the mountains became visible again, dark shapes against the starry sky. I said I loved how the squalls blew down the mountain, for it reminded me of being at sea. Evy, my wife, said she loved just being outdoors. Richard's wife, Chris, said she loved *everything* about the baths.

Richard claimed he was too hot, so he lowered himself, sputtering invective, down the cedar ladder into the cold plunge. He remained there some long moments, trying to stay down longer than I had and still appear unaffected by the shockingly cold water. When he could stand it no more, he shot out like a breaching beluga and eased back into the hot pool.

"Hey, why don't we live like this?" asked Richard when he'd recovered from the shock.

"Why indeed?" rejoined my wife. "You're the plumber. You build one, and we'll help you use it."

And so we passed the evening, talking, soaking, occasionally torturing ourselves with the cold plunge just to feel the renewed heat of the hot pool, marveling at the stars and snow.

This was one of the grimmer aspects of our "This Old House" Santa Fe, New Mexico, project. We were at a place called Ten Thousand Waves, an establishment designed like a Japanese *onsen*, or traditional bath, which offered a dozen hot tubs, all outdoors, each situated so that it was completely private. When you checked in at the front desk, you were issued your own robe and slippers, which you could don in the meticulously clean, wood-paneled dressing room, and then you made your way to your pool along the graveled pathways lined with low bamboo fences and piñon trees.

Once we had discovered the place, we returned often, and for all of us, creating an environment like this in our own homes became a great fantasy. Such a fantasy is not that farfetched these days, when whirlpool tubs, saunas, and even lap pools are commonplace in the bathrooms of American homes.

Twenty years ago, American Standard, a manufacturer of plumbing fixtures, produced a modest promotional film entitled *That Room Down the Hall,* aimed at getting people to pay more attention to the bath, then a small, utilitarian chamber where one performed one's ablutions as quickly as possible and left, in order to give the next person a chance. Few people in the early 1970s built the bathroom any larger than it had to be, and still fewer outfitted it so generously that they wanted to show it off.

An *onsen* is a traditional Japanese public bath centered around natural springs. This elegant rendition in a hotel outside Tokyo is finished in cypress and marble. The scrubbing stations at right are for thoroughly cleansing oneself before sliding into the serene pool.

Czar Nicholas I's bath in the "Cottage" in St. Petersburg. The oversize washstand of ash and marble is the centerpiece for a comfortably appointed dressing room. The curtain in the foreground hides a spiral staircase, down which the czar's advisers would come to deliver the morning reports while he dressed.

These days, it would be hard to sell a public relations department on the necessity of making such a film. The home magazines and the general media are filled with visions of opulent or merely handsome baths, and most of us are aware of the amazing profusion of bathroom fittings and accessories. As American families are increasingly focused on their homes, the bath has become the place to soothe away the tensions of the day and indulge the senses.

●

In most old houses — say, anywhere from fifty to a hundred and fifty years old — the rooms that require the most work are kitchens and bathrooms. Typically, the last time these rooms were remodeled was in the 1940s or '50s, and the renovations tended to be

utilitarian at best. When I renovated my own house, an 1836 Colonial Revival on Boston's North Shore, I was faced with three 1940s-era bathrooms. My first thought was to keep them and just do a cosmetic makeover, but my wife insisted they must go. I gutted and replaced all three, and even added a fourth in the master suite, with a walk-in shower and a whirlpool that overlooks the garden.

The main living areas of our house cover about three thousand square feet — ample, but not palatial. Not counting our offices, located in the attic, we have four bedrooms. One might, therefore, consider four bathrooms a bit excessive, as I did when faced with the renovation. But now, almost five years later, I'm glad my wife was insistent. With the number of people in our house — visiting friends and neighbors, children, weekend guests, and colleagues stopping by the office — four bathrooms turns out to be just enough. Even at that, my wife still complains bitterly every time she finds *my* whiskers in *her* sink, and mutters that we should have still more bathrooms, or at least more sinks.

Bathroom as shipboard stateroom. Maplewood vanity with undermounted china lavatories, 1940s-style faucets, a bank of drawers — and, of course, sweeping views of the harbor make this master bath feel as if it is aboard one of the grand ocean liners of a past era.

The claw-foot tub, pedestal sink, and painted pineboard wall and ceiling evoke a sense of *déjà vu* in this summer home on Martha's Vineyard.

In a way she's right: double sinks in the master bathroom would ease congestion during the morning rush when we're both getting ready at the same time. And it would be very handy if the downstairs half-bath were larger, had more storage, and included a shower to accommodate the odd extra guest or my rinsing off after one of my late fall or early spring windsurfing sessions. The fact is, we, like most Americans, spend the vast majority of our time at home in either the kitchen, the bedroom, or the bathroom. When renovating or building new construction, taking some time to focus on the purpose of each bath and the interaction among all the baths in the house will greatly enhance the quality of your home.

●

At "This Old House," we've been redoing obsolete bathrooms and building new ones since 1979, when we took on our first project, in the Dorchester section of Boston. In renovating that rundown Victorian, we built two new bathrooms back to back on the second floor, incorporating a washer and dryer into one of them. We installed

Sunlight washes the slate walls and floor of this spacious sit-down steam room/shower that verges on the outdoors.

what turned out to be a somewhat expensive sloped tile floor in the laundry bathroom (for the inevitable day when the washer breaks and spills gallons of sudsy water), but generally we tried to keep tight control over our expenses by specifying standard fixtures of good quality yet moderate cost.

We had a bigger bath budget in our next project, the conversion of the H. H. Richardson–designed Bigelow mansion in Newton, Massachusetts, into condominiums, and we used it, in part, to install higher cost fixtures such as whirlpools and bidets. We shoe-horned powder rooms into handy but inconspicuous places on the ground floors of some units and created what we thought was a sleek and sexy black tile bathroom in one of the larger units. (We're not so enthusiastic about the black tile now — like a shiny black car, it shows every speck of dust and dirt.)

Since then we've completed forty additional bathrooms, using just about every type of appliance, fixture, and material on the market: pedestal sinks in our Reading, Massachusetts, New Orleans, and Wayland, Massachusetts, projects, a shower for two at the Weatherbee Farm, a marble-veneered whirlpool/shower stall in Santa Fe, New Mexico (in that bath, the walls were radiant heated to keep the bathers nice and warm), steambath and exercise facilities in our Arlington, Massachusetts, project, and a barrier-free bath, equipped to accommodate guests with disabilities in our Lexington, Massachusetts, bed-and-breakfast project.

We can't say we've seen it all, but we have done bathrooms all over the United States and even in Britain, too. We've built tiny baths in which every inch and every dollar counted, more lavish master suites where comfort rather than budget was the rule, and many bathrooms in between. In all the "This Old House" bathrooms, and in our own renovations — for all of us on the show, master carpenter Norm Abram, producer-director Russ Morash, plumbing and heating expert Richard Trethewey, and I work on our own houses and have to face these issues as homeowners — we have tried to make choices that would instruct us and our viewers about materials, products, installation and building procedures, design and aesthetic considerations, and budget matters. Working with a succession of homeowners, each with his or her own goals, has kept us flexible in our attitude toward the bath; the challenge we always face at "This Old House" is to find the solution that fits the house, the budget, and our homeowner's way of living. Although we have preferences — which we will not hesitate to tell you about in the chapters to follow — there is no one "best" material for a shower or tub, no one "best" design. Bathrooms must suit a variety of needs and tastes; the one that best addresses yours is the room that is right for you.

We hope this book represents the collective wisdom of the "This Old House" team. Norm Abram, of course, has been with the show from the very beginning, infusing his knowledge and great good sense into everything we've ever done, as has Richard Trethewey, who has made it his personal mission to stay current on the very latest plumbing and heating technology and hardware. The folks at Design Associates, Architects, especially Jock Gifford and

Christopher Dallmus, have done excellent design work on "This Old House" projects, as has kitchen and bath designer Glenn Berger of Acton Woodworks. No list of acknowledgment would be complete without further mention of executive producer and director Russ Morash, who dreamed-up "This Old House" in 1979 and is still the creative mind behind the show. Not content to direct Norm and our other "This Old House" experts during the week, Russ usually spends his weekends working on his own 1851 farmhouse outside Boston. His latest project? To renovate the downstairs utility bathroom.

My cowriter, Phil Langdon, who also covers architectural topics for *The Atlantic Monthly*, interviewed the "This Old House" experts as well as other specialists dealing with every aspect of bath design and renovation. He revisited many of the baths we renovated to see how our work stood the test of time. Our homeowners showed no reluctance in telling him about what worked and what they would do differently.

Bathroom renovation is complex, though not as complex as the kitchen. Most bath remodels will not expand the limits of the current room, so there is minimal carpentry involved. There may be some reframing to move a window or door or to install a skylight, and there will certainly be some electrical and heating work. But in the construction process, much of the expense is in the plumbing.

Bathroom renovation is mostly a design, fixtures, and materials story. Whereas we usually have only one kitchen in our home,

(Above left) Russ Morash, producer of "This Old House," sets up a shot on the Wayland project with Rich Trethewey and cameraman Dick Holden.

(Top, right) Richard Trethewey, heating and plumbing expert for "This Old House."

(Above) Norm Abram, master carpenter for "This Old House."

(Right) Postmodern asceticism realized in stainless steel, brushed plaster, and marble. Notice the lighting behind frosted glass panels engineered into the mirror.

which must perform many functions, not the least of which is to pre-pare meals, we typically have two or more baths, each serving one principal function — a powder room, a children's bath, or a master bath. A large part of preparing for a bath renovation has to do with defining the "functional specifications" of the room, which is to say what it must do, who will use it, and where it fits in the usage patterns of the household. When renovating some types of bathrooms, such as the master bath or the guest bath, you will want to think through the larger design implications, circulation patterns, and privacy issues that arise when these baths are being used. A second part of the design process is evaluating the various fixtures and materials now available for the bath, on the basis of utility, beauty, and cost.

We have organized this book to look at the design and material considerations of the various types of bathrooms, from the utilitarian to the ultimate. We start with general remarks about design and planning, then move right into discussion of the various bathrooms: from simple powder rooms to utility baths, children's baths, barrier-free baths, exercise baths, master baths, and master suites. In each chapter we discuss the design, construction, and materials considerations that apply to that particular bath.

As you browse through the text and illustrations, keep in mind that no matter what type of bath you are building or renovating, the difference between an adequate bathroom renovation and a really outstanding job often lies in the design of the details and how well they are integrated into the room. Even though you may only have the money for a modest makeover of the family bath, you may pick up some good ideas in the other chapters — it doesn't cost a cent to dream. It is likely that your bath will have to perform the functions of several of our somewhat idealized bathrooms — the powder room might have to double as a washroom or the family bath as the romantic spa (after the kids are asleep, of course) — so we recommend looking through the whole book before settling down to read in earnest the chapter concerning your project.

One of the great freedoms we've enjoyed in the course of "This Old House" renovations has been the opportunity to try innovative products, materials, and methods. In the pages that follow, we present a realistic appraisal of our experiences — successful and otherwise. Frankly, I wish I'd read it *before* I renovated my own bathrooms!

●

This Old House Bathrooms

A Guide to Design and Renovation

Design and Planning

Bathrooms are design intensive. From the division of the larger space to the selection of materials and the execution of the small details; this elegant design gets it right. Note the raised ceiling, which defines the central axis from the more intimate areas — WC, shower, and tub — that are placed in the wings. For good detailing, check the placement of windows, lighting fixtures, and ventilation. For good choice of materials, note the granite tile, the marble counter and backsplash that tie the space together, and the mahogany bowfront vanity, which is the focal point of the room.

In many ways, the toughest part of most bath remodels is just getting started. Chances are, you've become increasingly dissatisfied with your current bath (which is why you're reading this book) either because it is too small, poorly designed, or merely outdated, ugly, or downright squalid. And chances are you've been collecting ideas about what you would like to put in its place. Remodeling can be as simple as replacing fixtures, flooring, and finishes, or as complex as gutting a section of the house to make room for a large master bath with whirlpool and skylights. Costs can range from several thousand to tens of thousands of dollars. So the first order of business is to step away from the glossy photographs of luxury bathrooms you've been clipping from magazines and become coldly realistic about what you want your renovation to achieve.

Don't delude yourselves, as I usually do; just about any bath renovation is a pain in the neck. It will disrupt the flow of life around the house, it's messy and complicated, and it can involve every building trade from carpenters, electricians, and plumbers to tile setters and painters. The general rule in renovation is that costs go up, not down; so your bath will probably be more expensive than you initially bargained for.

Therefore, the first question to ask yourself is how long you plan to stay in your house. Answer realistically, because it is the first major fork in your decision strategy.

If your answer is three years or less we recommend you do the minimum in the bath, although there is no hard-and-fast rule on this point. A survey of real estate brokers by *Remodeling* magazine indicates that after a bathroom has been remodeled, the owners recoup at resale an average of 73 to 88 percent of the cost. Adding a second full bathroom within the existing floor area (as opposed to adding on) pays off a little better, recovering 85 to 98 percent of the investment. In neither case does the homeowner break even, although a renovated bath may help to sell the house faster, which may have a dollar value, especially in a slow real estate market. Note that these are national estimates for projects well thought-out and constructed and appropriate both to the house and to the neighborhood. If the neighborhood is not in great demand or the project

contains some flaws, the payback could be considerably lower.

If the bathroom in question is less than ten years old, remodeling will probably do little for your home's resale value, unless the bath has taken an unusual amount of abuse. In general, a ten-year-old bathroom shouldn't be terribly out of fashion and should still be in pretty decent shape. Brokers say the biggest potential for payback is in houses twenty-five years or older that are in top-notch condition except for the bathroom. If your bathroom fixtures look tired and outdated, and the tile is a color that has long ago gone out of style, you might consider replacing them; but don't underestimate the effect of new paint, wallpaper, lighting, linen, rugs, and accessories. I have participated in enough architectural photo shoots to appreciate the difference accessorizing a room can make in the final "look."

When sprucing up a bathroom for resale, our advice is to stick with standard fixtures in white or off-white and in general to avoid any flamboyant or unusual decorating scheme that may turn off a prospective buyer. You might also talk to some local real estate agents about what their buyers are looking for in your neighbor-

A cosmetic renovation entailing new brass fittings on the old tub and sink, a new mirror instead of the old medicine chest, and a handmade mosaic tile floor. The design goal here was maximum style at minimum expense.

(Left) "Styling" makes this simple bath an elegant one. New paint and a beautiful shower curtain do all the work.

(Right) A new floor, a coat of green paint, and a whimsical vanity completely transformed this bath.

hood. This can help in making appropriate choices, which may vary from one community to another.

In some very old homes, the condition of the bathrooms and plumbing may be so bad that the house will not fare well in an engineering survey. In this case, you may need to modernize the baths because a deteriorated bathroom may discourage potential buyers from even making an offer. *Remodeling* magazine observes that most homebuyers see a bathroom remodel as a major project they prefer completed *before* they buy the house.

If you do not plan to sell and are renovating for the long term, you can follow your own tastes and not worry so much about the current real estate market. Ironically, these types of renovations are often more difficult, because the universe of possibilities is suddenly expanded (cost efficiency *does* simplify the process), but ultimately they can be very creative in terms of use of space and materials and certainly can be accomplished at reasonable cost. The success of your new bathroom is largely determined before the first swing of the sledgehammer to demolish the old. It all has to do with design.

Predesign

At "This Old House," we think a full-blown bathroom remodel should serve you well for at least the next ten to fifteen years. You may change the flooring, fixtures and counters, but the basic design should endure. Over that time it will pay for itself in function, beauty, and enjoyment, and in all probability will enhance the resale value of your house. For a bathroom to fulfill these expectations, it must be thoroughly thought out and well designed.

Architect Jock Gifford feels the design phase of a successful bathroom remodel will take *at least* as long as the construction phase. Yes, this adds to the cost, but paying a design professional what he or she is worth to help you develop and build the right design for your house and family is far cheaper than building the wrong bathroom.

If design is critical, "predesign" is just as critical. There are two exercises I do before I even contact an architect or designer.

This sophisticated bath glows like a fine china lacquer bowl. Mirrors lining the wall and vanity base reflect the black marble, black toilet and bidet, and brass fixtures and electronics panel.

Chances are, if you've been thinking about your bathroom for some time, you have already done both, perhaps without realizing it.

Critique

One is to critique your existing bathroom. This exercise simply involves trying to articulate what *you* think is "right" and "wrong" with the room. I like to go into my bathroom and patiently "look." I try to identify what bothers me about the room as it exists, and what I could do to change it. Mentally, I alter the room's colors, flooring materials, and lighting scheme; I move tub, sink, and toilet, and install or remove windows, skylights, and doors. Often, I have found, what bothers me is not one major flaw, but a lot of small things that cumulatively ruin the feeling or the function of the room — undersize mirrors, no place to put shaving cream and razor, poor lighting, and so on. Sometimes, after a careful analysis like this, you can figure out ways to drastically improve your bathroom without major work, thereby saving yourself a lot of time, money, and aggravation.

You might wonder why you should bother with this exercise if you're already sure you will renovate the bathroom. The answer is that you can know if your new design will work only if you can articulate why your current design doesn't.

Dreaming

The next part of predesign is dreaming. Again, you've probably done a good deal of this already, which is partly why you're reading this book. Once I've determined the bathroom is too dilapidated or outmoded to save, I jump to the other extreme and try to visualize what I would do if I could do anything, cost no object. Move the bathroom, extend it into a greenhouse addition in the backyard, take over the master bedroom — *anything.* The purpose of this exercise is to articulate what you most love. Dream big — it's costing you only a little time and maybe some paper on which to sketch your ideas. You can scale back later when you must.

Dreaming is critical to the design phase because it helps to break up any preconceived limitations you may have placed on your project. You may find that you *can* have your dream master bedroom–bath suite, but phased in over several years. You should spend your first money knocking down walls, putting on an addition, rearranging doors and windows — whatever it takes to organize the space for a sound design that will last. You can always install the whirlpool or sauna as budget permits.

People tend to think that major restructuring of space — moving partitions, bearing walls, windows, doors — is prohibitively expensive. It may be more or less expensive depending on your house's structural characteristics. You may well find (as I often

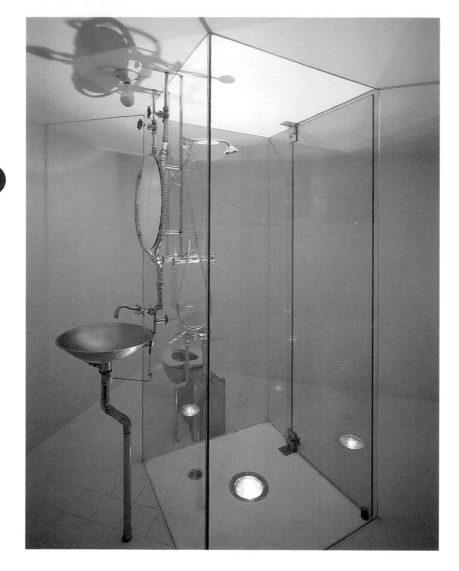

Bathroom turned upside down — including the lighting, which shines up from below!

have) that you can rearrange your space in a major way and still maintain your budget by scaling back elsewhere: installing counters of plastic laminate instead of marble, and using a standard tub instead of a whirlpool.

A Design Notebook

Most designers and architects suggest that early in the predesign process you begin a file or a loose-leaf notebook in which to organize your thoughts about your bathroom project. Make a list of all the things you like and dislike about your current bathroom, and about other bathrooms you've used. You can also keep a wish list, and a collection of photographs from the home magazines of features and details you really like.

Much of what you first put down in this notebook may seem obvious or trivial. Don't abandon it, for you'll find as you continue

Design Points

Whether you are hiring an architect or plan to tackle part or all of your bath design and renovation yourself, here are some factors to consider from the very outset of the design phase:

Comfortable Heights

Your bathroom will be much more comfortable to use if it is well matched to your height and physical abilities. While "standard" height of counters and lavatories is 32 inches, this dimension originated at a time when Americans were shorter, and now some people find this height uncomfortably low for using the sink. Many of our "This Old House" homeowners have specified their vanity counters at 36 inches, the standard kitchen counter height, but you can specify any height that you find comfortable. If the bath will be used by both tall and short people, consider installing two lavatories at different heights, or one lavatory at a mutually agreeable height with the countertop at two different levels. As in all design considerations, we would avoid features so idiosyncratic as to discourage a future buyer.

Seating

Consider whether you need seating other than the edge of the tub or the toilet lid. Perhaps a stool or chair could be planned along a run of counter facing a mirror, or a built-in seat engineered into the bathroom's casework. Built-ins are especially useful for dressing and undressing small children. They may also provide a way to hide a radiator or to increase storage.

Compartmentalization

In a bathroom for more than one person, it may make sense to put the toilet in a separate compartment or even follow the European practice of placing the WC in a separate room. For the more public bathrooms in the house, this arrangement increases the number of people who can use the facilities at the same time. In a master bath, each partner has more privacy (a condition one increasingly values as one gets older). The drawback is that such an arrangement takes up additional space. Another compartmentalization tactic is to locate a vanity in the bedroom or makeup area outside the bath.

Sharing the Bath

Before bathrooms became so numerous, it was not unusual to equip them with two doors, one serving one bedroom, the other communicating with a hall or another bedroom. This is still a useful design strategy especially in an occasionally used guest room or study that shares a powder room or family bath.

Focal Points

The bathroom, like the kitchen, is a room in which the "furniture" is permanently installed, so considerable attention should be given to what fixtures get placed in the room's natural focal points and in sight lines from adjoining rooms. It would make sense to place a handsome pedestal sink or whirlpool bath in such a dominant position, instead of the toilet or bidet. Focal points can be highlighted with accent lighting.

Safety

People will use a bathroom at all hours of the day and night, while half-asleep, naked and wet from the shower or bath, and without their spectacles or contact lenses. Not surprisingly, more accidents occur in this room than in any other, so safety and user-friendly ergonomics should be a primary design objective.

At "This Old House," we are reluctant to introduce any change of level in the bath that might increase the chance of a fall. Similarly, we do not like steps at the foot of bathtubs or whirlpools, even though there are photographs of such in this book. If steps are unavoidable, they should be well lit, have a nonskid surface, and be accompanied by grab bars or railings.

Shower and bath controls should be placed where they can be adjusted *before* getting into the bath or shower. We also recommend installing a pressure balancing antiscald valve (even if such is not required by local plumbing codes) to insure against scalding from abrupt changes in temperature when water is drawn elsewhere in the house.

Shower doors should slide in tracks or open outward — never inward; such an arrangement would impede the rescue of anyone who might fall or collapse there.

Lighting throughout the bath should be of good intensity with low glare. The floor should be slip resistant.

the exercise that it is an excellent way of identifying all the aspects of the bathroom that matter to you, not just the two or three factors that led you to consider remodeling in the first place. The biggest problems, and therefore the biggest potential sources of improvement, are often easy to spot. Smaller problems and less obvious potential improvements tend to be overlooked unless you approach the predesign phase systematically, which this exercise helps you to do. The great advantage of list making is that it encourages a comprehensive approach to understanding what is to be achieved. After going through the process, you'll be able to define your objectives with greater thoroughness and precision. The more clearly you understand what you want, the more effectively you will communicate with a designer or contractor.

Also compile a list of *specifications,* or "specs" — any items you already know you want: two sinks, a tub big enough for two, a particular type of showerhead or style of toilet. Some of your requirements may not yet be clear in your own mind, but starting a list will help get them out on the table to be discussed.

An efficient way to help work up your spec list is to pay a visit to a local plumbing showroom, which has on display many types of fixtures, faucets, vanities, and accessories. There is one showroom in the Boston area in which all the fixtures are hooked up, so you can actually see how they operate. To assess whirlpools and spas, customers are encouraged to come dressed in their bathing suits and jump right in!

This bathroom in a Maine farmhouse was enlarged with a bump-out niche for the tub, a tactic that kept the room's original character and nearly doubled the apparent space without a major renovation.

The plumbing itself makes the major design statement in this bath, with a custom spigot and exposed plastic drain hose. Notice the resonant curves: hose, sink, back-splash, and mirrors.

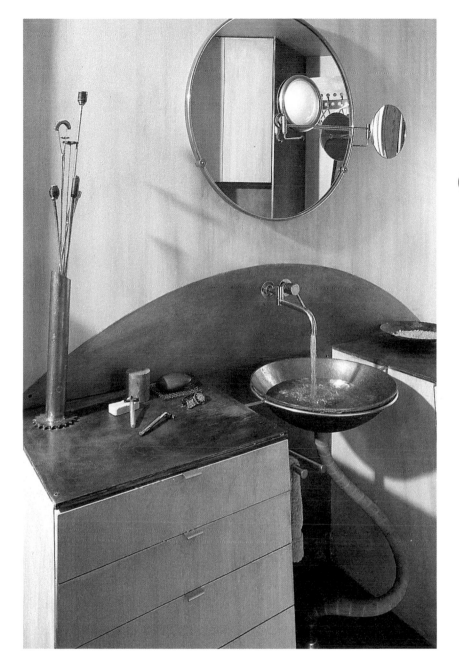

Jock Gifford of Design Associates suggests that all household members, even children, review the lists and add their own requirements or desires. This will make it easier for all of you to get together and talk about features that the bathroom should possess. Each person who will be using the bath can offer suggestions from his or her own perspective, and giving everyone a say will help determine the project's true priorities.

As part of your preparation, browse through a number of books and magazines collecting clippings and photocopies of fix-

13

This hewn-log ranch house in Utah feels every inch the gentleman's shooting lodge it is. Notice the "found object" hooks for towels and washcloths.

tures, fittings, materials, layouts, and any other features that appeal to you. Again, you might think this exercise seems trivial, but Chris Dallmus at Design Associates says: "The pictures in the notebook help to give the architect a sense of the client that he or she may not be able to articulate. Inevitably, the clippings will reveal the client's likes and dislikes, sense of space and color — in a word, style." Kitchen and bath designer Glenn Berger agrees: "If you take your project to a designer and say, 'Do my bath,' that is an endless realm of choices. By helping to define your requirements, tastes, and style, you narrow his field of inquiry, and help him to focus right away." Going through your clippings with your designer will help him or her zero in on the appropriate colors, materials, spatial organization, and other elements of your bath.

With practice, you can become adept at evaluating the bathrooms you see in magazines and books. Photographs (including the ones in this book) often glamorize a bathroom by lighting it very selectively or decorating it with unique accessories. A vase of exotic flowers by the bathtub, a lush fern hanging beneath a skylight, or a rustic old wooden chest can give the photo a magical allure, but if you're trying to visualize how the room would actually work, you may need to omit the flowers, delete the beautiful linens, and focus instead on the room's basic components and floor plan. Looking at photographs with a critical eye not only helps you extract the essential information from them, but will also help build your skills for when you must evaluate your own prospective floor plans.

The whole point of this predesign phase is to get ideas suspended in your mind so that you can make the most efficient use of your architect or designer in the next phase — design.

Using a Design Professional

There is no such thing as the "ultimate" bath, especially in a rehab. Nor is committing the time and money to the pursuit of the ultimate a particularly good way to come up with the robust. "Successful bath renovations are inspired by limitations," says Jock Gifford, "of budget, space, design, and construction timetables." These constraints inspire the designer, contractor, and client to focus on the essential and leave aside the superfluous.

In planning and executing my own renovations, I've always had to work hard to get the greatest effect for the least money. The best way to achieve this is through elegant design. I use the word in the scientific sense: finding the simplest solution to the problem. If you clearly define the problem in the predesign phase, you can go about solving it in the design phase.

The design problem here was to renovate the bath without expanding the room or departing from the apartment's 1920s style. The tile floor replaces a worn one of the same style; the console sink evokes the original but is larger; and the new faucets and hardware are selected from the many reproduction models available today. A narrow whirlpool replaces the claw-foot tub.

At "This Old House," we would not consider doing a renovation without an architect or design professional whose work we know and admire. We've often trusted the professional's advice and been glad we did. We like the fact that architects work directly for their client and earn no commissions by pushing particular products. This gives them a certain freedom to recommend whatever they think best suits the project.

Architects can bring a comprehensive approach to design problems; they are generalists trained in all phases of design and construction. Some architects are also trained in project management and have experience in bidding and contracting, scheduling, budgeting, and managing relations with subcontractors. I am a typical owner-builder, and even with my interest in design I tend to see design changes mostly in terms of how much they will cost and how hard they will be to build. The architects' job has always been to argue for the design itself. In this way, a healthy tension has developed between us that has benefited the project.

I personally would not undertake *any* renovation without consulting an architect or designer. I've done so in the past and regretted it. Even if I *know* what I want, I always run my plans past an architect to get his or her input. Each time I have, the architect has improved the design to the extent that I considered the fee some of the best money I spent.

•

If we've convinced you to use a design professional, the next logical question is where to find one. Not every architectural firm is available to handle bath projects. Large firms tend to be geared toward large commercial projects, leaving residential work to smaller firms with lower overhead. Yet, not every small firm can do a good job. Finding the right architect for you is a bit like finding a psychiatrist or a barber: the one who is right for you is the one who is right for you. Peruse books and periodicals for designs that catch your eye. Architects who are just beginning to win recognition are not necessarily more expensive than those who are lesser known.

Perhaps the best method of finding an architect is to ask friends and business associates for recommendations. Homeowners who have just completed a bath remodel are usually more than happy to discuss their projects with you. You can get lots of free advice, some of it valuable. Talk to the architects they recommend, look at their work, and make your own judgments.

No matter how the architect comes to your attention, be sure to talk to his or her other clients and visit the jobs. This will give you a lot of insight into his or her style and capabilities. It is important that the personal chemistry is good with your prospective architect. If, after the initial meeting (for which there should be no charge), you feel you cannot work together, keep looking, because any bath remodel, especially a large one, will likely be stressful. You need to know you can communicate very clearly and immediately with your design professional.

Architects typically work under one of three different methods of payment. Probably the most straightforward is an *hourly fee,* which can range upwards from $50 per hour. This approach may be preferable if you are just exploring the possibility of renovation or if you want to do some of the design work yourself. Under this scheme, you can save yourself some money by handling your own research on fixtures, fittings, materials, and accessories. Many architects are happy to delegate such tasks, reserving their own time for the more creative and specialized work of design.

(Left) A lot of bathroom in this awkward loft space. The glass block surround solved the problem of how to place and make watertight the shower. The countertop is slate, with undermounted lavatories.

(Right) Vivid colors and an eclectic mix of old and new give this modest bath the sense of a Paris brasserie. The absence of a shower curtain keeps the room open.

A second approach is a flat fee, termed a *stipulated sum.* It has the advantage of limiting the client's expenditure for design, and is suited for clients and architects who have previously worked together or for clients experienced with renovation. In a stipulated sum arrangement, the architect may limit the number of site visits or hours he or she will devote to the project.

A variation on this strategy, one often used by Chris Dallmus, is to work within a "not to exceed" budget. The architect agrees to do the design, working drawings, on-site supervision, and contract administration, or any subset of these tasks, for a figure not to exceed a mutually agreed-upon sum. Chris likes this arrangement since both client and architect are constantly comparing what's been spent to the ultimate sum and measuring that against the work yet to be done. This consciousness of the limits reinforces effective communication and allows for constant adjustment of their contractual agreement.

The third method is a percentage of the construction cost. This percentage may vary with the scope of the job: typically, the smaller the job, the larger the percentage. Percentages can range from 10 to 20 percent. We, along with many architects, consider the percentage method undesirable, as it penalizes the architect for finding ways to economize.

If your renovation is a relatively modest one — if you are planning to stay within the confines of the existing room, contemplate no alteration of windows or addition of skylights, or merely want to update an existing bathroom with new fixtures and finishes — a bath designer may give you all the design assistance you need. Glenn

Berger says that the line between architect and bath designer has become increasingly fuzzy in the past several years. With the expanding number of new bath products on the market, design and renovation have increasingly become a matter of knowing which sinks, tubs, faucets, cabinets, whirlpools, and steam baths are out there, and how all these components mesh together in the tight real estate of the bathroom. A big part of bath design is mixing and matching manufactured items, and bath designers tend to be specialists in this regard.

Bath designers are often resident at kitchen and bath stores, firms that design these rooms, sell the components, and may also perform or manage the renovation. Often, a bath designer working with a showroom will make an initial site visit and work up a preliminary sketch for anywhere from $50 to $150, and produce actual drawings for $300 to $500. This sum is often termed a *design retainer*. You can take the drawings elsewhere for bid, in which case the design retainer becomes a *design fee*; if you buy the components from the firm or have them do the work, the retainer will be subtracted from your final bill.

Berger's firm will do the design work for free, as long as he then gets to supply the materials and perform the work. Alternately, he will design and manage the job for a 10 to 20 percent markup on materials and labor. For straight design work, he charges $50 to $75 per hour. (If you have trouble finding bath designers, the National Kitchen and Bath Association, at 124 Main St., Hackettstown, NJ 07840, can provide a free directory of member firms that design, supply, and install residential bathrooms.)

●

How do you choose between an architect and a bath designer? While there is no absolute rule, we would say that in a renovation that involves structural work or reconfiguring a portion of the house, such as creating a master bedroom–bath suite, or a guest suite that can double as an area for a nanny or an elderly parent, one is advised to seek an architect, who can help define and work through the larger issues. In our opinion, using bath designer or showroom personnel to design your project is advisable only if it is a fairly simple one. By the nature of their business, bath stores run the risk of looking at the bathroom mainly from the standpoint of what fixtures the homeowners want and where they want them, possibly overlooking imaginative solutions regarding light, views, circulation, and other architectural considerations. But some homeowners, having lived with and thought about their baths, know just what they want done. For those of you in this situation, a bath designer or bath store may be an excellent choice.

The shower also serves as an entrance to the terrace. The roll-down screen on the door lets in light but not the neighbors' eyes.

We should add that an experienced bath designer may also be superb at space planning and may know the details of bathroom design, equipment, and finishes quite well.

The Evolution of a Plan

Once you have decided on a design professional, he or she will come to your site, measure the room, and ask you a variety of questions. Some of our "This Old House" homeowners have been surprised at the depth and breadth of the questions our design professionals have asked. These questions are critical, for the designer is doing, in perhaps a more structured and thorough way, what we strive for in predesign. The objective of the first meeting is to understand all the problems of the current bathroom and the re-

This sophisticated bath packs a lot of well-designed features. Light from an atrium-lit hallway filters into the shower; a bench in the shower allows for seating when the steam feature is on. Stereo speakers are fitted into the soffit; a phone is nearby. The sculptural pendant lights add humor while solving a restriction in the location of the electrical boxes. Walls are tiled in brush-hammered marble — the step before the marble gets polished.

quirements of the new room. The designer will undoubtedly look through your file or notebook and take some notes on it.

The result of this meeting will be a set of preliminary plan alternates, which he or she will then present at the next meeting for discussion and review. It is possible one of the designs will be just what you want, but more likely, each plan will contain elements you like and elements you don't. This is the beginning of a "design dialogue" between you and your designer, which will help to advance and refine the design.

Very few people can evaluate a floor plan right on the spot, so you may want to review the plans and then take them away with you for further study. In order to fully understand them, we suggest you tape the floor plans, full size, on the garage or basement floor. Mark the positions of the sinks, toilet, bath, and shower and go through the motions of using the room. If two of you will be using it simultaneously — say, in the morning when getting ready for work — you both should go through the exercise together. Find out if there's enough room for both to wash up at the same time or if one will be able to apply makeup while the other is shaving. Keep your notebook with you during these experiments and jot down your observations so you can share them with your designer at your next meeting.

Be sure you fully understand what you like and dislike about every plan, for as we said before, you can know what is right for you only if you know what is *not* right and why. I enjoy the design process tremendously. I particularly like the interchange of ideas with the architect. Do not be afraid to suggest radically different solutions to your design problems, even if they might sound silly to you at first. The nature of the design process is very much one of inspiration followed by the hard work of realization. You may, in fact, have a very silly idea that contains the seed of a wonderful new plan. In the best of all situations, the creative energy flows between designer and clients in such a way that the new space is a synthesis of everyone's energies and ideas.

Also keep in mind that design is not reserved for professionals only. If you want to try your hand at designing your bathroom, we encourage you to do so. If you have a personal computer, you might buy an interior design software kit that lets you look at different floor plans with relative ease. You can also buy a kit of cardboard cutouts of toilet, lavatory, bathtub, and other components. Lay them on graph paper to test different possibilities.

Before you build, we encourage you to pay an architect to critique your plans. I once did the design for a renovation of the first floor of a small weekend house in the country that my wife had

Most old houses were built with too few bathrooms, and those constructed before the days of indoor plumbing had none at all. Thus, finding the square footage to install a new bath or enlarge and renovate an old one becomes the first problem facing the renovator. But square footage matters little if the location of the bath does not relate sensibly to the house it serves.

In order to illustrate how the process of overall space planning might evolve, we created this imaginary project. It is based on a real house, the front rooms of which date from the late 1700s, and was therefore built without a bath (one was added to the second floor in the 1940s). Readers of *This Old House Kitchens* will recognize it as the same concept house we used there; indeed, the kitchen and family room reflect our design. We explore alternate room layouts in the appropriate chapters.

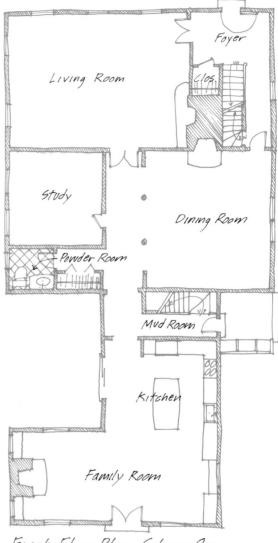

First Floor Plan - Scheme One

Our first objective was to get a powder room on the first floor. In looking for space to commandeer, we gravitated toward the study, a room whose utility would not be seriously compromised by further reduction in floor space, and which is centrally located and just off the house's main axis. The closet provides additional storage while helping to give it privacy.

bought when I was off in the Pacific doing research for my book *The Last Navigator.* I had drawn the kitchen, living room, and bath/ laundry room very neatly, complete with measurements and elevations, and proudly presented my completed plans to our friend and designer Marilyn Ruben. Without comment, she took a piece of tracing paper, laid it over the floor plan, and so thoroughly revised my drawings that nothing was left unchanged. She moved the kitchen to the living room, opened it onto the dining room, and moved the living room to the kitchen. I was gasping; after all that work, she was throwing it away! Yet because I'd done the work, I understood

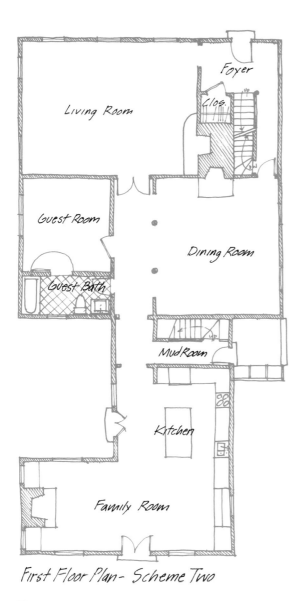

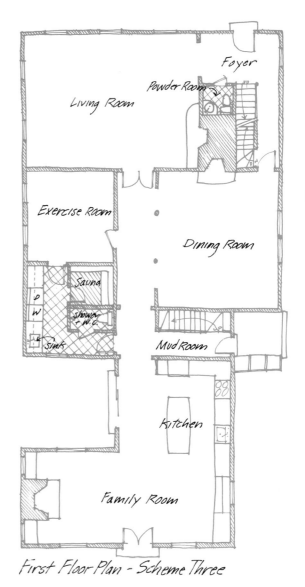

First Floor Plan - Scheme Two

First Floor Plan - Scheme Three

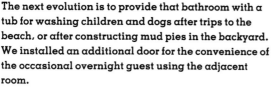

The next evolution is to provide that bathroom with a tub for washing children and dogs after trips to the beach, or after constructing mud pies in the backyard. We installed an additional door for the convenience of the occasional overnight guest using the adjacent room.

For additional space, we pushed out the south wall of the room to install a combination exercise/utility bath, with washer, dryer, and sauna. Also in this scheme, a small powder room, intended for the use of guests, replaces the closet at the bottom of the front stairs.

why she was right. I entered into her aesthetic in a way I never could have had I not immersed myself in the problem and groped for my own solution. I have had similar occurrences with other designers since. Ironically, these designs, the ones I had initially got "wrong," are the most satisfying to me.

Test any designs you think are winners, no matter who has generated them, on the garage floor. We also encourage you to build a simple model out of cardboard or foamboard to see how the plan looks in three dimensions, with walls, ceiling, and windows. Making models is a quick and easy way to troubleshoot a prospec-

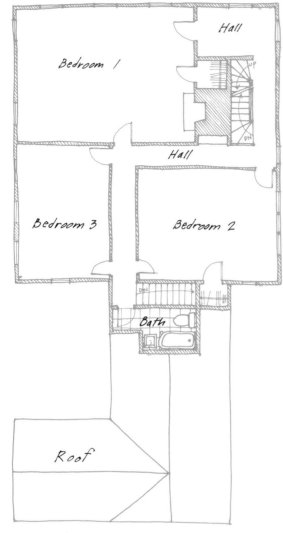

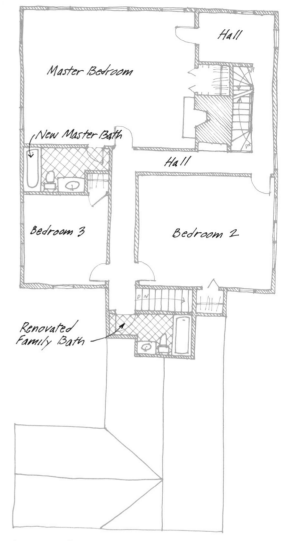

Second Floor - Existing Conditions

Second Floor - Scheme One

The second floor had one small bathroom, which had been added under the eaves of the back ell long after the house was built.

The idea here was to add a second bath for the least possible cost. We installed a modest master bath in space stolen from the second bedroom. The existing bath was updated with new fixtures and finishes within the existing footprint.

tive design or refine an already evolved one. Some foamboard and a few hours of your time is cheap; an imperfectly conceived plan realized in wood, plaster, and tile is expensive.

●

Once you have finalized a design, your architect or bath designer will need to produce a full set of "working drawings," usually larger-scale plans that describe the work in detail. Such detailed drawings will help you to get more accurate bids from contractors and tradesmen, and in so doing they may even pay for themselves. In drafting the plans, the architect must think in terms of construc-

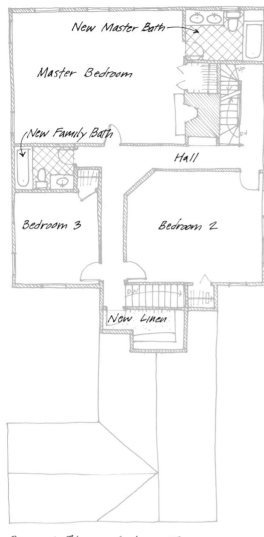

Second Floor - Scheme Two

Second Floor - Scheme Three

Instead of renovating the existing bath, we gutted it and converted it into much-needed linen storage. We placed the family bath in the center of the bedroom group and wedged a master bath into the former second-floor stair landing.

Here we enlarged the front bedroom by utilizing the second-floor landing, and raised the roof over the back ell to make room for a large master bath and dressing area.

tion, not just design, so potential problems can often be solved right there on the drawing board, instead of at the last minute in the field.

"I've seen the situation too often," Chris Dallmus says, "in which we do a bare set of design drawings and go right into construction, since the client wants to skip the working drawings. Then we get a call from the builder on the job site scratching his head trying to figure out how to build what we've drawn. We end up doing crisis intervention instead of design; coming up with a solution — probably not the best — just to keep the project moving."

●

Getting the Job Built

Great old fixtures can sometimes be found in salvage shops, such as this one in Maine, which scavenge them from a variety of sources, clean them up, and recycle them. Keep in mind, however, that adapting an old plumbing fixture for use with modern hardware may end up costing more than a new reproduction fixture.

Finding the Help You Need

With a set of working drawings, you're ready to get the job built. You have several options open to you, in addition to doing the work yourself. The one you choose will largely depend on the scope of your project.

If you plan to make extensive structural modifications or add windows, skylights, or greenhouse bays, the logical choice would be a contractor who is experienced in such work. For smaller projects, you might award the contract to a kitchen and bath center that can offer a turnkey job. If any problems occur with workmanship, materials, or components, the kitchen and bath center bears the full responsibility. Another possibility is a plumbing contractor. My longtime friend and plumber, Rich Bilo, does complete remodeling of both kitchens and bathrooms. He is a skilled mechanic who works tile, wood, and solid surfacing with equal facility. If you can locate such a man in your area, you will benefit from the fact that the responsible party in such a plumbing-intensive job as the bathroom is the plumber himself.

Yet a fourth choice is a qualified carpenter capable of performing most of the tasks involved in remodeling a bathroom. A carpenter may not be able to do all the work, for in some jurisdictions, electrical and plumbing work can be performed only by a licensed journeyman. In any case, you do not want such work performed by amateurs (with the possible exception of yourself).

If you've successfully worked with a particular contractor, plumbing contractor, kitchen and bath center, or carpenter (I'll refer to all as "contractor" in the following discussion), you may want to hire that person again and not bother soliciting bids from competing contractors. If you feel comfortable committing yourself to a contractor before he or she has enough information to quote a definite price, you can cut the length of time you'll have to wait for construction to begin. The contractor can schedule your job before drawings and specifications have been prepared as long as she or he knows that the detailed information that is needed will arrive in time. Since, except in the slowest of times, good contractors are usually booked months in advance, this arrangement can speed the process along. This tactic is useful on a big job, such as a major master bath renovation, but may not be appropriate for a small job.

If you're new in town or don't have a favorite contractor, it's wise to solicit more than one bid on your project. Request several references from each contractor and call the former customers to find out how satisfactorily the contractor performed. In soliciting bids, be sure that the bid specifications are as complete as possible. "You've got to make sure everybody is bidding on exactly the same project," Norm Abram stresses. "The more decisions you make about materials, finishes, and detailing before calling for bids, the easier it will be to select a contractor, and the fewer problems you will have once he begins the job." If you ask for bids before specifying all the materials to be used, it's common for the contractor to include an *allowance* in the contract, a line item of a specific sum or price per square foot to cover items yet to be selected. But a contract with many allowances is a weak document and doesn't give you much certainty about the eventual cost of the project.

Upon receiving bids from contractors, ask for an explanation of anything that is unclear. If there are wide disparities in bids, ask the higher bidders why their prices are high. It may turn out they are offering superior workmanship or materials. Beware of the very low bid. In tough financial times especially, some contractors will bid a job below their break-even point just to secure the contract. Then, they will try to turn a profit by increasing the price of changes and extras, or by lowering quality. Such a "low" bid may well turn out to be more expensive than those that were fairly calculated. If a contractor knows he is losing money on your project, he has little incentive to finish it, or if he does, he has little incentive to do a good job.

There is some efficiency in using an architect and a contractor who have worked together, because they are familiar with each other's methods. The architect will know how much detail he has to include in his plans for the contractor, and the contractor will have a surer sense of how the architect wants his plans built.

An alternative is to act as your own general contractor. This may cut costs 10 to 25 percent, although predicting such cost savings is a bit dicey. Bathrooms are small, intense jobs, and as Russell Morash, Sr., father of the "This Old House" director and a professional contractor, used to say: "It's the corners that cost money." Alas, the bathroom is full of corners.

For you to fill the role of general contractor successfully, we recommend that you seek out the best tradesmen in your area and pay them what they are worth. You may pay more, but you will save money in the long run. My experience has been that a topflight

28

In such a design-intensive space as a bathroom, careful execution of the details is critical. Here, the tiling, woodwork and shuttered windows, and rounded marble counter are beautifully fitted. Quality of workmanship is particularly important. Notice how the tiles are evenly matched and cut on either side of the window, and around the corners and the mirror.

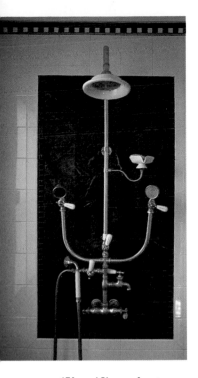

(Above) If you plan to use recycled, nonstandard, or foreign plumbing fixtures, like this Portuguese shower spray, make sure they're on the job site before rough-in so your plumber can anticipate any unusual plumbing requirements.

(Left) The owner of this bathroom acted as her own general contractor. Rather than replacing all the old tiles, she had Zolotone, an industrial-strength spray-on coating, applied, creating a wonderful marbling effect. The retro-style floor is new.

craftsman who feels trusted and respected on the job will look for ways to make the project better. Often, he will make small improvements as a matter of course, at no additional cost.

You might also take on some of the work yourself. Framing a simple bath might involve only a few dozen studs, making the project small and contained enough to be handled by someone good with his or her hands. You might well tackle the tiling, essentially a simple job, which, with enough time and patience (and availing yourself of the plentiful books and manuals on the subject), you could do very successfully. Hanging your own gypsum board is also within the range of many homeowners' skills. For rough plumbing, the average homeowner would be advised to use a professional, even if this is not required by local code. However, you could probably handle the final plumbing: installation of the toilet, lavatory, and shower controls. You can also take on the demolition. Many homeowners derive a demonic pleasure from blasting out the old plaster, tile, and fixtures. If a contractor can begin with a neatly gutted room, it is that much less he has to do and that much less he will charge you for.

You can also realize savings by taking on the chore of chasing down materials: tile, plumbing fixtures, flooring, specialty faucets, and so on. Many home centers sell items at low prices, sometimes lower than the contractor's wholesale price, so you may save a fair bit by shopping around. But many contractors are reluctant to enter into such an arrangement with a homeowner, and for some good reasons. First, the contractor may lose a 15 to 20 percent markup on the fixtures and materials, a source of his or her job profit. Or he or she may fear you will obtain the wrong products, in incorrect quantities, or fail to have the materials on hand when they are needed, leaving the workmen idle and costing the contractor money.

If you do take on this job, treat it seriously. Be ready to drop everything to get supplies when the contractor needs them. "It can't be done at the homeowner's convenience," says contractor Tom Silva, who with his brothers has done several projects for "This Old House." Some items must be ordered well in advance, so you must keep on top of your construction schedule.

Perhaps one of the best ways to be involved in your project is to hire a contractor or plumber and act as his helper. This way you benefit from his knowledge and experience (not to mention his tool kit) while saving what he would have to pay an assistant.

With all these caveats in mind, doing your own contracting can not only save you money but can give you the deep satisfaction of having built it yourself.

Contracts and Negotiations

When you get a bid from a contractor, the price specified generally remains available for only a limited time, commonly thirty to sixty days. If the contract is signed within that period, the price stands. If not, the contractor is under no obligation to stick to that price. Frequently, prices of materials change after the contract has been signed. When that occurs, the contractor is the one to absorb price increases or benefit from price reductions; you pay the contract price. If the contractor made mistakes in calculating his bid, this, too, should be his responsibility, not yours.

In many projects, especially in older homes, expect the unexpected. The crew opens up a wall, for instance, and finds rotten sills or floor joists that must be replaced. In old baths, the rot will probably be found around the toilet and tub. Complications not readily anticipated are usually your responsibility unless the contract specifies otherwise. Some contractors (including Norm) will not bid a job until all the demolition is completed and any structural damage assessed. Some contractors will show a "contingency" line item in their bid to cover such eventualities. The contingency may range from 10 to 25 percent, depending on his assessment of the condition of the building. You should be reassured rather than alarmed upon noticing this item, because it shows the contractor is realistic. Indeed, you should plan your own contingency of 25 to 30 percent, as projects tend to get bigger once under way, rather than smaller.

Here in New England, where the average home is fairly old (my own was built in 1836), nothing is plumb or square, and you never know what disaster lies behind that horsehair plaster wall. When working on old buildings, my friend John Ashworth, an architect/builder, figures his bids as tightly as possible, then *doubles* that figure. In my experience working on my own houses, that's just about right.

Prior to signing the contract, it's a good idea to discuss with the contractor what problems he will cover at no extra cost and what problems will be your responsibility. Make no assumptions about this; the contractor may have different expectations from those we've laid out here. The more you anticipate beforehand, the more smoothly the project will proceed.

If you request a change after bids have been figured and contracts signed, you, naturally, must pay for it. "But the cost may be substantial," warns Norm. "Extras are always marked up more than things that were in the contract at the start." Some of this may be understandable; from the time the contractor started looking at your plans to bid them, he planned to build the job a certain way —

This vaulted ceiling heightens the impression of space and light, even though the room is barely wide enough to accommodate the claw-foot tub. The tub itself provides a focal point, while the color ties it in with the accent tiles on the walls and floor.

your change may have ramifications you don't readily perceive. This underscores the importance of good planning and design.

Whenever a change is made, it should be written as a *change order* and attached to the contract. Amended drawings should be attached to become part of the contract as well. Typically, the change order is a simple form that tells what aspect of the job is being changed at what cost. The contract needs to be kept up to date to avoid misunderstandings. This will minimize the possibility of both a nasty shock of a big bill at the end of the project and of arguments between parties with conflicting recollections of what was agreed upon.

Once the job is under way, some homeowners are afraid to take issues directly to the contractor, approaching instead a member of the construction crew. This is unfair to the worker and undermines the contractor's chain of command. Confront the contractor directly (and in private) with problems and give him or her an opportunity to solve them. In general (but unfortunately not always), if you treat a contractor with respect, he or she will do the same for you.

Office supply stores sell standard forms for contracts. Many of these are one-page documents too short to cover many of the situations that arise in the course of a large or complicated renovation. More detailed contracts for use with an architect or contractor are available for a small charge from the American Institute of Architects' headquarters in Washington, D.C., and from AIA chapters throughout the country. Although regarded as cumbersome by many contractors, standard AIA contracts may be beneficial for the homeowner, since they provide more control over the contractor than does the typical one-page stationery store document. Before you sign *any* contract, we advise you to have it examined by a qualified attorney.

Many contracts list a date by which the contract is to be completed. Penalties for failure to meet deadlines can be specified in the agreement. Some contractors will start your job and then jump to another project before returning to finish your job. Sometimes this is done in order to keep cash flow going by doing just enough to collect the payments on each job. Other times, though, a contractor must operate this way for overall efficiency, by keeping ahead of his subcontractors on all his projects. If you expect your contractor to work your job every day until completion, discuss this during your negotiations and note your understanding in the contract.

Payments for a construction or renovation project can be scheduled in whatever way you and your contractor agree. There is no universal method, although it is not uncommon for one-third of the project's cost to be due upon signing the contract, the second third upon completion of rough framing, plumbing, and electrical work, and the final third upon completion. It is common to withhold 10 percent until the contractor has completed all the items on the punchlist — those details yet to be attended to. "By holding back 10 percent, you're holding the contractor's profit," Norm explains. "If the holdback is much less than 10 percent, some contractors may be slow to 'punch out.' "

In larger renovation projects, it is common to bill the client weekly; the bill, of course, being accompanied by an accounting of work performed. This improves cash flow for the contractor and keeps the client abreast of the cost of the project. No matter how the job is billed, the contract should specify what remedy you have if the contractor does not complete the job satisfactorily.

Old brass and copper pipes from our Wayland project go into the dumpster. Sometimes copper and brass have salvage value.

Most states require a three-day cooling-off period, during which you may back out of the contract with no penalty. If you don't know whether there is such a provision in your state, call your consumer affairs office in your municipal, county, or state government, or talk to your attorney.

As important as a contract may be, your biggest asset is your good relationship with your contractor. Norm has done many jobs — especially complex renovations — on the basis of a letter of agreement rather than a fully detailed contract. Like Norm, many contractors are craftsmen who build as much for the satisfaction of a job well done as for their paycheck. If you're lucky enough to find such an individual, it's important to recognize and respect his (or her) talent and abilities. A dedicated craftsman will often go out of his way to improve the job.

●

As I've said, bathroom renovation is more a matter of design options than of the mechanical work itself. The plumbing, heating, electrical work, and carpentry will be fundamentally similar from one type of bath to another. Before we go on to the chapters dealing with the various design and materials issues associated with the different types of baths, let's review some of the basics.

Demolition

Bathrooms are tightly organized spaces constructed with many heavy, durable materials: tile floors, counters, and shower stalls; cast-iron tubs and waste lines; ceramic sinks and toilets. When these components were installed, they were installed for good, a fact that often makes renovating a bath an all-or-nothing proposition.

Bathrooms built prior to 1950 often had ceramic tile floors laid using the *mortar bed,* or mud-job, method, whereby a layer of mortar 1 to 2 inches thick was poured on the subfloor and the tiles then pressed into it. A tile job of this kind can take year after year of hard use, but can also pose quite an obstacle at renovation time. You could try to work around it, patching the floor as necessary, but getting matching replacement tiles would be no easy matter. As a result, there's often no alternative but to rip out all the old tiles and mortar.

Similarly, old pipes and their attachments can sometimes be re-used, especially cast-iron waste stacks and vents, into which your plumber can tie his new waste lines. Cast-iron elbows and unions *can* be altered, but doing so requires arcane materials such as oakum caulking and molten lead, installed by a trained mechanic. Most plumbing will be cheaper to replace than to fiddle

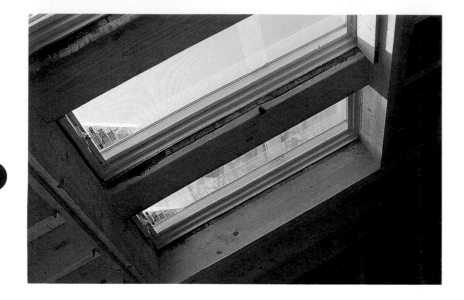

36

with. After years in place, nuts, bolts, and collars will have developed a coating of rust, and when subjected to pressure, the connections may break. Here in the Northeast, it is not uncommon to find old plumbing of fine threaded brass, which, though still serviceable, may crack or break at the connections when you try to patch into it. Some plumbers will want to replace *all* the brass piping rather than risk causing a leak in some hidden spot in the plumbing system. It is often faster and cheaper to plan on gutting the bathroom and starting from scratch than trying to work around old fixtures, floors, or plumbing in order to try to save a few bucks.

Demolition usually is carried out by the general contractor to ensure that construction work won't be delayed. If you choose to do the demolition, allow yourself plenty of time — holding up your contractor can be costly.

One of the toughest items to remove is a mud-job tile floor. In the first "This Old House" project, a Victorian in the Dorchester neighborhood of Boston, we used sledgehammers to break up the floor, a crude but effective method. These days you can rent an electric demolition hammer at your local rental store, which will speed things along.

Another difficult aspect of the demolition will be removing the old cast-iron tub. Ordinarily it is installed after the subfloor is in and the walls framed, but before any finishing work. Because the floor and walls are finished up to it, be prepared to see these surfaces ruined.

Framed master bath at our Wayland site.

If the bathroom contains some old work you would like to incorporate into the new bath, it's wise to remove and store those materials rather than leave them vulnerable to damage by the construction crew. A lot of rough treatment occurs during a renovation project, and it's easy for old work to be marred or ruined, even when the contractor is trying his or her best to keep it intact.

Some materials generated by the demolition, such as old radiators and brass and copper pipes, may have salvage value, but much of what's removed will simply have to be disposed of. In many communities, materials from a construction site are not accepted for routine municipal trash removal, so a dump truck or Dumpster is called for, thus adding to the cost of the project. Here in Boston, a big thirty-yard Dumpster can cost from $600 to $1,000 per load, depending on the final weight.

In the course of demolition it is not uncommon to discover hidden damage. In many of the old houses we've worked on, there originally was no shower, only a tub, with a window some distance above it. At some point a shower spigot may have been added and the walls tiled to form a surround. Often, the windows were left right in the middle of the shower wall, and over time, water splashing on

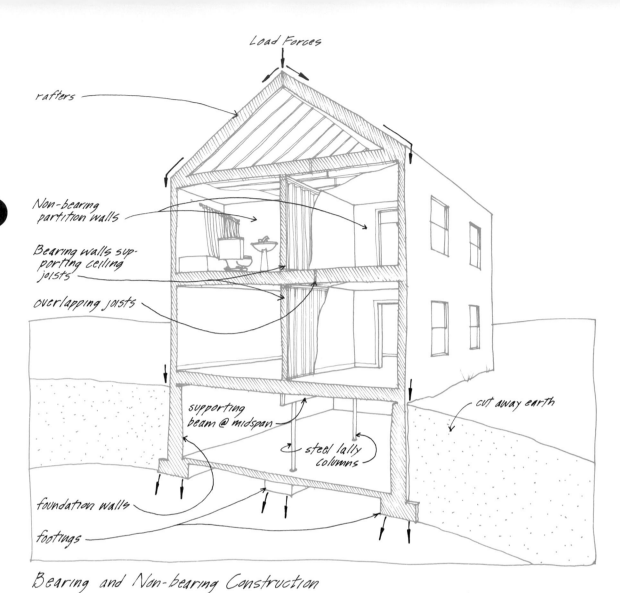

Load Forces

rafters

Non-bearing
partition walls

Bearing walls sup-
porting ceiling
joists

overlapping joists

supporting
beam @ midspan

cut away earth

steel lally
columns

foundation walls

footings

Bearing and Non-bearing Construction

Removing a Bearing Wall

"Today just about any wall can be removed if you take the right precautions," says Norm Abram. The load that's been carried by the studs can be carried instead by a beam. The beam will then pass its load to posts, beefed-up studs, or other framing members. Beams are available in steel, solid wood, or laminated wood. Steel is heavy and difficult to attach wood to, but it is very strong and thus can be sized to be easily concealed in a wall or ceiling. Steel is cheaper than wood, but more expensive to work with. Solid wood beams, on the other hand, need to be sized very large to carry a heavy load or accommodate a long span. Laminated beams are made of many thin layers of wood glued together. They are very strong yet can be worked with standard woodworking tools. Joists

can be attached directly to them via metal joist hangers. Laminated beams can be ordered in several wood species. They are *prestressed* — that is, the top curves upward to counteract the downward pressure of the building's weight — and come marked "top" and "bottom."

Before removing a bearing wall, it's necessary first to build temporary walls, or shoring, to carry the load until the new beam and other supporting members are installed. The size and placement of temporary walls as well as beams and other framing components should be specified by an experienced carpenter or structural engineer. If you're tackling a project like this for the first time, seek the advice of a professional.

the exposed wood deteriorated it and perhaps seeped down into the wall behind the tile. If the window is in bad shape, this may be the first sign of moisture damage inside the wall. Whether or not rot has developed, the window should be moved out of any wet area in the new design. Water and humidity are enemies of a house's structural integrity.

Similarly, water could have been seeping into any junctures of horizontal and vertical surfaces — between tub and floor, the tub surround and the wall, or between the toilet and the floor. It's essential to investigate and repair any damage.

Another potential source of structural problems is, ironically, past plumbers. In the Dorchester and Wayland houses and in many old houses that were built in the days before indoor plumbing, bathrooms were installed in converted bedrooms. In the process of running pipes through floors and walls, the plumbers of yore often cut away framing members to the point that they failed. Damage of this kind may be impossible to detect before the demolition is under way. The only reliable procedure we know of is to have an expert inspect the framing after the walls or floors have been opened, and then beef them up if they're insufficient.

The need for a strong floor framing system is especially great near the bathtub. A cast-iron tub of average size can weigh 300 pounds. Filling it with water adds a few hundred more pounds. When the bather gets in, the total burden becomes quite a strain on the supporting structure, so it is only wise to look carefully for any hidden structural weakness.

Walls

While most baths are renovated within the existing walls of the room, some projects will call for expanding the bath's present boundaries to make space for a separate tub and shower, for double sinks, or simply for more floor space.

In order to move a wall, you must first determine if it is a *load-bearing* or a *non-load-bearing* (or *partition*) wall. A bearing wall is essential to the structural integrity of the building. A non-bearing wall simply divides space in the house. You can find out if the subject wall is bearing or nonbearing by examining floor or ceiling joists wherever they're exposed, such as in the basement or the attic. If the joists run parallel to the wall, chances are it is nonbearing; if the joists run perpendicular to the wall, it is probably bearing.

In a house that has neither basement nor attic, searching for the locations of joists is a bit more difficult. You can drive an eight-penny finishing nail into the ceiling at two-inch intervals (in some inconspicuous location) to locate the joists, or, better still, use an

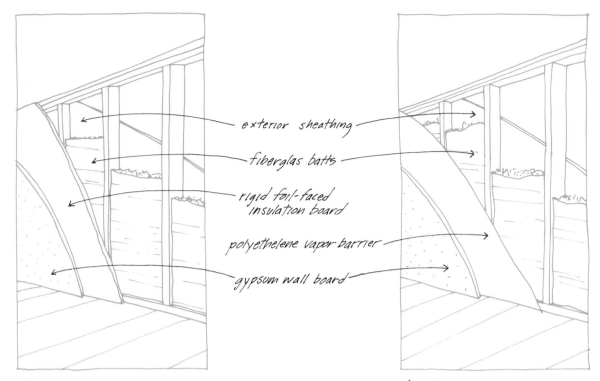

exterior sheathing

fiberglas batts

rigid foil-faced insulation board

polyethelene vapor barrier

gypsum wall board

Insulation option one   Insulation option two

electronic stud finder, commonly available at home centers and hardware stores.

Many people mistakenly assume that bearing walls cannot be altered or removed, but "making openings in bearing walls (except in multistory buildings) is not the big deal it once was," advises Norm. "With the variety of wood, steel, and laminated beams available today, just about any wall can be removed — with the proper precautions and engineering."

You do not need to suffer a confined bathroom just because your expansion plans would require moving a bearing wall. In general, the costs of structural framing is a small percentage of the total cost of the bathroom, the lion's share being composed of finish materials and labor.

Vapor Barriers and Insulation Systems
In winter, when you spend the whole day outside in boots, even well-insulated ones, your feet seem to get colder and colder as the day progresses. When you pull your boots off at night, your feet probably will be damp. This is from perspiration, and as your socks got more and more damp throughout the day, their insulation value was destroyed, resulting in cold feet.

The same occurs in a house. In winter, the interior is warm and moist, while outside it is cold and dry. It is a phenomenon of

On a "This Old House" project, foil-backed insulation is carefully installed. The strips should fit snugly on the top, bottom, and sides without compressing the loft of the glass fibers. I always wear gloves, a long-sleeved shirt, eye protection, and a good dust mask or respirator when installing insulation.

physics that environments want to equalize, and as the warm moist air from your house seeps through the exterior walls, the moisture condenses on the insulation, where, just as on your socks, it reduces its insulating value.

To counteract this, a vapor barrier should be installed between the interior wall surface and the wall studs. Commonly it is a sheet of polyethylene film that is glued and stapled to the studs, but foil-backed gypsum board and rigid, foil-backed insulation board can be used as well. If you are gutting your bathroom, installing a state-of-the-art insulation and vapor barrier system is worthwhile.

At "This Old House," our favorite method is to insulate the stud bays with unfaced batts of fiberglass insulation and then cover the walls and ceiling with rigid foil-backed polyisocyanurate foam board, sometimes called energy board. Not only will you benefit from the insulation in the walls themselves, but you will get an additional, unbroken blanket of insulation from the board. For an effective vapor barrier, all seams should be taped with aluminum tape, and electrical boxes and other breaks in the board caulked with a compatible foam.

Foil-backed energy board comes in ¾-inch and 1-inch thicknesses. However, some bathrooms may be so tight on space that you don't have room to install insulation board, in which case we recommend insulating the walls with fiberglass as described above and stretching a sheet of six-mil polyethylene over the walls and ceiling. All breaks in the vapor barrier should be sealed with latex caulking.

Rough Plumbing

The average residential plumbing system can be broken down into three subsystems: *supply lines,* which deliver hot and cold water throughout the house; *waste lines,* which convey contaminated water to the sewer or septic system; and *vent lines,* which allow gases in the waste lines to escape to the atmosphere and permit proper drainage.

In the old days, the plumbing system was frequently a hindrance to the remodeler's plans. Cast-iron waste lines could be modified only with difficulty, using oakum and hot lead. Today, most jurisdictions allow the use of polyvinyl chloride (PVC) for waste and vent lines. PVC is light and strong and can be quickly and easily cut and glued together with simple tools. The interior of the pipe is more slippery than cast iron, a trait that helps prevent clogging.

Whereas, in the old days, supply lines were made of galvanized iron or brass, the pieces of which had to be laboriously threaded and screwed together, copper tubing is now universally used. Copper can be cut and soldered together quickly and reliably. Increasingly, we see the use of plastic supply lines as well, either polybutlyene (PB) or crosslinked polyethylene (PEX), which can be snaked like electrical wiring throughout the house. The salient point here is that new piping materials allow tremendous freedom in locating the bathroom and its fixtures.

Ideally, plumbing communicating between floors should run in a *plumbing chase,* an uninterrupted shaft from basement to attic that allows for the systematic design and installation of vent, waste, and supply lines as well as electrical, telephone, and television cabling. Unfortunately, it's often difficult to find a good spot for a plumbing chase in an old house, but we think it's worth some effort to engineer one. Had I taken the time to do this in my own house, subsequent plumbing and electrical installations would have been much easier.

In a climate with cold winters, such as New England's, it is vital that the water supply lines be kept out of exterior walls. When "This Old House" added a second-story bathroom to a house in Boston's northern suburbs, we ran the pipes through a first-floor closet to the second floor rather than risk the pipe in the exterior walls. It doesn't take much air infiltration to make a copper pipe freeze and burst during a January cold snap. If you must install supply lines in

(Top) PVC drain connected to cast iron via a stainless-steel and rubber clamp. Cast iron is used here to cut down on noise transmission (the PVC in the ceiling will be insulated). The orange-striped tubes conduct radiant heat to the master bedroom above.

(Bottom) PVC waste and copper feed lines for the first-floor powder room at our Wayland site.

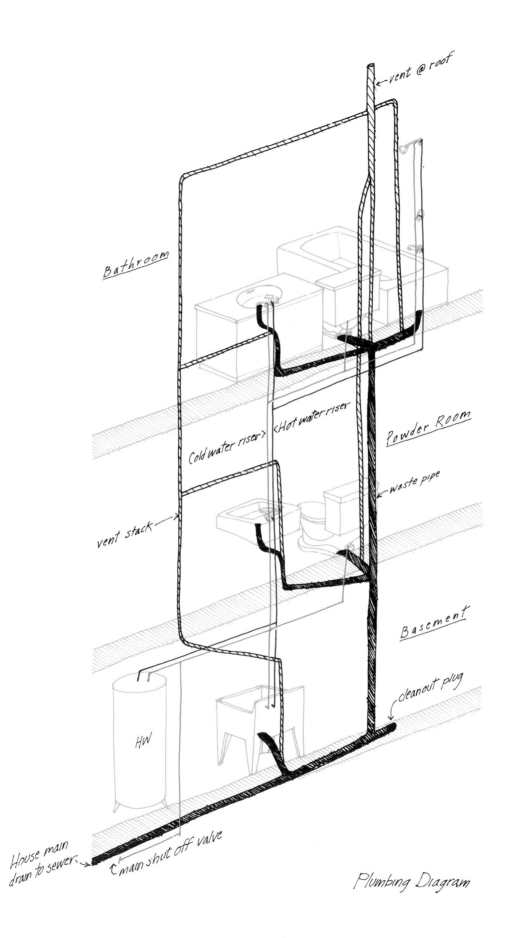

vent @ roof

Bathroom

Cold water riser> <Hot water riser

Powder Room

vent stack →

waste pipe

Basement

cleanout plug

HW

House main
drain to sewer →

main shut off valve

Plumbing Diagram

These "roughed-in" bathrooms in a "This Old House" project are piped back-to-back to keep all the plumbing in a single "wet wall." Note the copper feed pipes and the PVC waste and vent lines are capped and ready for pressure testing.

an exterior wall, they should be heavily wrapped in insulation. In extreme cases, the pipes can be wrapped with heat cables.

If your project is a minor renovation, you may try to limit your plumbing costs by keeping the new bathtub, lavatory, and toilet in their original locations. But before letting this determine your whole layout, consider that plumbing often can be moved a few feet without too much additional cost or difficulty. Keep in mind, though, that a bath layout dictated solely by the existing plumbing may be at odds with the room's ultimate workability. Although plumbing is not inexpensive, it is not so dear that it should prevent you from getting the design you want. We think it is worth trying to fit the plumbing costs into the budget rather than make do with an awkward layout.

One way to plan for a more elaborate bathroom without building it all at once is to have the plumber run rough plumbing to the future sites of, say, a walk-in shower, steam bath, or sauna area. He can then cap, or "stub out," those pipes, making it easy and relatively inexpensive to install those features at some future date.

Electrical Rough-in

Electrical design may not be the most glamorous element of your project, but, like plumbing, the electrical rough-in comes early in the renovation process and requires careful planning.

Electrical work is tightly governed by standards laid out in the National Electrical Code, which is embodied in federal law. State or local codes may impose additional requirements. Local building inspectors are responsible for determining if the work meets the standards. The system of inspections and approvals by local building inspectors sees to it that an electrical installation is safe and functional. It is foolish to try to skirt code requirements or avoid inspection. For the price of a permit, the code operates to your benefit, ruling out substandard work.

Electricity reaches your house from the local power grid via a *service line,* which is made up of heavy strands of copper or aluminum wire. The power passes through the electrical meter and then to a main circuit breaker and an electrical panel equipped with circuit breakers or, in older systems, fuses. The panel distributes power throughout the house in *circuits* that carry either 110 or 220 volts. In the United States and Canada, 110 volts is standard for lighting, outlets, and most household appliances (overseas, 220 volts is standard), but electrical devices requiring a lot of power, such as electric hot water heaters, clothes dryers, ranges, and water pumps, run more efficiently on 220 volts.

Whereas *voltage* is roughly analogous to the *pressure* of water in a hose, *amperage* is analogous to the *volume* of the flow. The total volume of electricity a house can draw from the grid is regulated by the main circuit breaker (or fuse). It used to be that 60 amps was considered an adequate capacity for homes, and in many older houses you still see what is called a 60-amp service panel — a main disconnect switch that will allow only 60 amps to flow into the house. Now, with the increased number and variety of electrical appliances in the average home, 100 amps is considered minimum and it is not uncommon to see a 200-amp service panel.

Electricity is distributed throughout the house in circuits — loops of wire that originate at a circuit breaker in the service panel and deliver electricity to a number of lights or outlets. The thicker the

(Left) When thinking through your electrical plan, don't forget to wire for all those necessary bathroom appliances. The mirror-mounted car stereo AM/FM cassette player (speaker over the shower) lets you shave while checking the traffic for the morning commute. As whimsical as the radio seems, it does require the installation of a 12-volt transformer, which means planning (and, incidently, renders the unit safe to operate while at the sink).

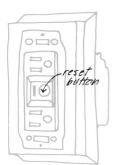

Receptacle GFI

Ground fault protection for one outlet.

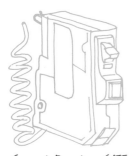

Circuit Breaker GFI

Ground fault protection for a large appliance such as a sauna heater or whirlpool

individual wire, the greater the amperage it can handle. If a wire's rated amperage is exceeded, it will overheat, with the risk of fire. Thus, each circuit is protected by a circuit breaker, which, just like the main service disconnect, will shut the circuit down if the electrical demand is too great. A common 110-volt circuit for lights and outlets is usually rated at 15 amps.

●

The typical bathroom has fairly simple electrical requirements: lights, fan, and a single outlet in the vicinity of the sink. Thus the room can be fed by a single circuit. If you are planning a more elaborate bath with whirlpool, steam bath, or sauna, your electrical requirements will be greater.

Code requires that all outlets within six feet of a sink be of a *ground fault interrupter* (or GFI) type. This is a device that senses a "ground fault," or short circuit, and will shut the circuit down to prevent electrical shock. These devices work very well, and it makes no sense to try to save money by skirting the code and not installing them. The best testimony to their effectiveness, and to the seriousness of electrical shock, is to be found at the ends of many tradesmen's extension cords, where they have installed electrical boxes containing GFIs to protect them and their employees from electrocution. Standard practice in a typical bathroom is to run the lights and fan through the GFI, thereby adding protection to all electrical devices in the room.

If your bath contains old wiring, we tend to accept the electrician's judgment as to whether it needs to be replaced or not. Old wiring should pose no safety problem as long as the insulating sheath is intact and the connections are still strong.

Older houses were commonly wired with armored cable, a flexible cable composed of two rubber-insulated wires bound together with tape and enclosed in a conduit of interlocking spirals of steel strand. Your electrician can judge whether the cladding is in satisfactory condition and if the insulation around the wires themselves is still intact. If the armored cable appears at all deteriorated, most electricians prefer to replace it, for a short circuit within can make the metal cladding the conductor!

The original method of residential electrical wiring — much older than armored cable — is called *knob and tube.* The electricians of that earlier era would run individual strands of insulated wire throughout the house, suspending them on ceramic insulators, or knobs. Where the wire had to penetrate a beam, joist, or plate, it would run through a ceramic tube. If knob and tube wiring is still in good shape, it is perfectly safe — electricians claim that it is even safer than modern wiring, as the two halves of the circuit, the posi-

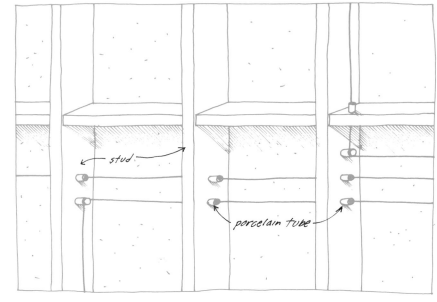

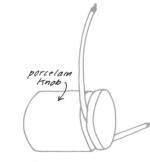

Old Style Knob and Tube

stud

porcelain tube

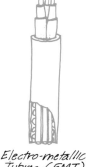

Electro-metallic Tubing (EMT)

tive and the neutral, are kept separate. I had an old Greek Revival house built in 1846 that was wired with knob and tube. When I suggested it should be replaced, my electrician would have none of it. Again, a qualified electrician can determine if the knob-and-tube writing in your home is still safe.

●

If you are planning a sophisticated, multicircuit lighting system, you may want low-voltage wiring for that purpose. This type of lighting requires 12 or 24 volts instead of the usual 110 volts, which simplifies the installation of the wiring to the extent that connections do not have to be housed in metal or plastic boxes. However, low-voltage systems also require special transformers and switches, so such a system must be carefully planned and specified in advance.

●

Non-metallic Sheathed Cable

We cannot overemphasize the importance of a methodical review — well before the electrical work actually starts — of all the electrical devices you want in your new bath. When the walls are open, wiring is cheap and easy. After the walls are closed in, it's harder and more expensive.

      If you are doing a high-end bath with all the bells and whistles, don't forget to wire for telephone, or even cable television and a sound system. Even in a modest master bath, a telephone by the toilet (as one sees in hotel rooms) is a real convenience, and wiring for it during renovation adds little to the expense.

Armored Cable

duplex outlet

outlet with ground fault interrupter

42" high outlet

recessed light

wall washer light

closet/valance light

single pole switch

telephone outlet

heat lamp

fan

smoke detector

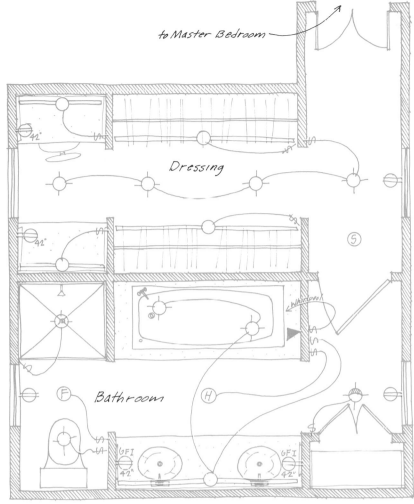

to Master Bedroom

Dressing

Whirlpool

Bathroom

Electrical Layout and Symbol Key

It might not be one of the most glamorous aspects of bath renovation, but study your architect's or designer's electrical plan carefully. Making changes before or during electrical rough-in is easy and cheap — making them after the walls are sealed is tough and expensive.

Lighting

Lighting the bath requires careful thought before construction begins, since you can't add plug-in lamps afterward, as in a living room or a bedroom. Lights are most commonly installed on the ceiling for general illumination, and above or on both sides of the mirror. Light fixtures may also be needed in a toilet compartment, in a shower stall, and above the tub, particularly if you like to read while soaking.

Opinion varies as to what kinds of lighting installations are most desirable. When there's a need for plentiful, shadowless illumination (as in a utility bath), it's not a bad idea to install fluorescent fixtures. They can even be recessed in the ceiling and covered with translucent plastic panels to diffuse the light. Fluorescent lights are the most efficient type of lighting in terms of electrical consumption,

and they give off much less heat than incandescent bulbs. When fluorescent fixtures are used in the bathroom, install "warm white" lamps, which render color more accurately and are, therefore, more complimentary to skin tones than the standard, industrial "cool white" tubes.

But even "warm white" lamps tend to pull out the blue tones in the face and skin — so you can give yourself a real shock in the morning when those bags under the eyes pop right out at you. Not surprisingly, most designers (not to mention homeowners) find this very unappealing and, therefore, specify incandescent lighting, often in the form of theatrical-type makeup lights.

Designers also find unappealing the idea of any fixture exposed in the center of the ceiling. If there are enough fixtures elsewhere in the room, such as above the lavatory and tub, there may be sufficient spill-off light for general illumination, making a ceiling fixture unnecessary. In higher-end bathrooms, where budget is not a big concern, you may want to devise a lighting plan that includes all three types of lighting: *decorative* (spotlighting a plant or an attractive object, for instance), *task* (for places where a man shaves or a woman applies makeup), and *general* lighting.

Inventive and unusual sources of illumination: Ample natural light comes from the window between the double sinks (check the reflection in the mirror), the skylight, and the porthole on the far wall; artificial illumination comes from lights in the skylight bay, heat lamps above the shower entrance, and fluorescents mounted on the mirrors. Notice also the beautiful teakwood vanities, the marble floor, and the brass sinks. . . . Looking around the corner into the tubroom, one finds both a skylight and a window.

(Left) Theatrical lighting, which itself is theater in this sculptural medicine shelf of brushed stainless steel and glass. Low-wattage fluorescents to the left cast light into the shower, while those behind the frosted glass above and below the mirror cast light on one's face.

(Right) Crisp design and a blending of ceramic materials take advantage of this subtropical location. Note the frosted-glass window and shower door, the use of glass block for the partition, and lights under the window and beside the door.

One of the basic rules in lighting is that you want to see the effect of the light but not its source. For this reason, recessed fixtures, or "cans," set into the ceiling can be very handsome in a bath, as can lighting installed in a soffit or box above the main mirror and concealed with a baffle or diffuser. In designing such a fixture, beware of placing it so that the light casts shadows on a man's neck while he is shaving or a woman's neck while she is applying makeup. Some designers place lighting along the sides of the mirror, which fully illuminates the neck. When installing recessed fixtures in a cold ceiling — one that borders an unheated roof or attic — special care must be taken to place the fixture on the "warm" side of the vapor barrier and insulation. Accomplishing this may require insulated housings around the fixtures, or even a false ceiling beneath the insulated ceiling. So intractable was this problem in our Wayland, Massachusetts, project that contractor Tom Silva talked us out of using any recessed fixtures in the master bath.

When I renovated my own house, I confronted the prospect of gutting and renovating two full baths and a half-bath and building a master bedroom–bathroom suite in what was once an apartment. I let the function and relative importance of each bath

Mirrors increase the sense of space in a bathroom and help to distribute the light. One can see the lighting above and below the entablature reflected in this mirror.

determine the lighting plan. In the half-bath downstairs, I used a single incandescent fixture in the center of the ceiling. It is a plain, white-frosted globe covering two ordinary light bulbs. It was inexpensive (about $25, as I recall), is perfectly inoffensive looking, and gets the job done. In the guest bath and child's bath upstairs, I used a single strip of makeup lights above the lavatory mirrors as well as the same kind of globe I used in the half-bath for general illumination. In the master bath, I used recessed cans above the whirlpool, a recessed can in the shower, and makeup lighting above the mirror. If I had it to do over again, I would probably opt for fluorescent in the kids' bath (children tend to leave lights on) and some kind of side lighting around the mirror in the master bath, to eliminate shadows on the neck.

Old-fashioned radiators still do the trick, while adding personality to the room.

Heating

Heating a bathroom presents special challenges. The room is small and filled with cold, hard objects and surfaces — cast-iron bathtubs, ceramic sinks and toilets, and tile floors, shower stalls, and walls — and most, if not all, of the wall area is commanded by permanently placed fixtures, which means there is little space to mount heating devices. Yet it is the very room you use frequently when partially clothed or downright naked, usually wet, too. From a heating specialist's point of view, this is a bleak picture: plenty of heat loss from the cold, hard objects, no wall space on which to put heating elements, and wet, naked people wanting a warm room.

If your bath is cold now, the first thing to do before renovating is to find out why. The problem might be the result of poor win-

This hydronic towel
warmer adds a sculp-
tural element to the
room, in keeping with the
antique brass lights and
mirror. It can be regu-
lated with a valve. Ter-
razzo tiles line walls and
floor.

dows or a lack of insulation in the ceiling or walls. If so, replacing
the windows with new, high-performance units and adequately insulating walls and ceiling may be the cure. But attacking the
sources of heat loss may not be enough; the bath may need more
heat delivered to it.

Usually, whether you're renovating existing space or adding on, the most cost-effective solution is to adapt whatever heating
system the house already has. A forced hot air system can generally
be reworked or expanded to serve the new bath. Sheet-metal ductwork is inexpensive, and one of the advantages of a forced hot air
system is that the ductwork can serve a central air-conditioning system as well.

If you have a hot water system with radiators, your plumber

European hotels of the class I could never afford always provided dry, warm towels with towel warmers similar to this one. To solve the problem of insufficient heat in my own master bath, "This Old House" plumbing and heating expert Richard Trethewey urged me to install a wall-mounted radiator like this one. Hot water from the heating system circulates throughout the hollow bars, keeping both room and towels nice and warm. Radiant panels come in a variety of sizes and heat outputs, both in electric and hydronic versions. Electric models can be equipped with a timer. Hydronic models are compatible with most American forced hot water systems.

may be able to adapt it to serve the renovated bath. But, as long as you are renovating, look at the stylish radiators manufactured by Runtal, DeLonghi, and Acova. These can be ordered in various colors, and some models even double as towel warmers. I recently retrofitted my master bath with such a unit and am very pleased. The room is warm, and the towels dry quickly in the winter.

If you don't want the look or expense of a new radiator, baseboard hydronic units can be installed along any free wall, or even behind a vanity cabinet outfitted with an air channel to conduct cool air to the unit and warm air from it.

In desperation, you could install a kick-space heater. Commonly used in kitchens, these units are designed to fit in the void space under the cabinet bases. They operate similarly to your auto-

mobile's radiator: hot water circulates through a coil that gives off its heat to a fan-induced stream of air. The fan speed can be regulated with a rheostat. Kick-space heaters produce a lot of heat, but like a forced air system, the heat is emitted in the form of moving air. If you are standing there wet and naked in a 72-degree breeze, you are going to feel as if it is 55 degrees in the room.

One of the big advantages of a hot water system is that it can be zoned to send more heat into some rooms than others. If you're adding on a bath, especially a large master bath or exercise bath, you might consider placing it on an individual zone. The bath is one room you want to be warm at any time of the day or night.

The Cadillac of delivery systems for hydronic heat is radiant floor heat, a system whereby flexible plastic tubing is secured to the subfloor and a thin slab of lightweight concrete poured over it. A stone or ceramic tile finish floor is then laid on the concrete. Alternately, the tubing can be run between wooden sleepers over which is laid the finish floor of carpet, sheet vinyl, wood, or tile.

Although more expensive to install than other hydronic delivery systems, radiant floor heat offers tremendous comfort. A tile or stone floor heated to 85 to 90 degrees radiates gentle, steady warmth into the room, like a stone wall that has baked in the sun all day.

If the floor area is too small for an adequate radiated surface, it's possible to install the tubing in a wall. We tried this method in the master bathroom of our Santa Fe, New Mexico, adobe house with good results.

(Left) In our Wayland, Massachusetts, project, PEX tubing is fastened to the underside of the plywood subfloor to radiant-heat the master bath above. Tubing is held in place by the aluminum reflectors. Although this is new framing, the same technique can be used to retrofit an existing bath with radiant floor heat.

(Right) Cross-linked polyethelene (PEX) tubing snapped into the plastic grid underlayment in the master bathroom in our Santa Fe project. The spacing and pattern of the loops is designed according to the heating demands of the building. Notice the tubing is more closely spaced at the outside wall than in the interior. The next step is to cover the tubing and the blue plastic underlayment with a slurry of lightweight concrete.

This porthole is an eccentric but clever solution for ventilating the shower. A fan placed high on the wall takes over when it's too cold to open the port.

It is possible to retrofit radiant floor heat by attaching the plastic tubing beneath the subfloor. The tubes are covered with aluminum reflector panels and then insulated with foil-backed fiberglass, rigid foil-faced foam insulation or even foil-faced bubble insulation. For this type of retrofit, you must have access to the underneath of the subfloor, either in the basement or crawl space, or by tearing out the ceiling below the bath.

I retrofitted my kitchen with radiant floor heat and found that, with the associated piping manifolds and electronic control equipment, it is technically demanding to install and at least twice as expensive as other systems. Still, it is so comfortable that if I were building a new house in New England, or substantially renovating an old one, I would install radiant floor heat. Similarly, I would also use it if I was renovating or installing a large master bath, especially one with lots of tile or stone. Otherwise, I would either extend the forced hot air system or use a radiant panel radiator in conjunction with a standard hydronic heating system.

Ventilation

Moisture and odors must be expelled from the room, so another system you should think through in the design phase is ventilation. Usually this is done with some sort of an exhaust fan. Even though most building codes do not require the installation of one if the bathroom has a window, we still recommend it. Inexpensive models are available, either fan only or with built-in light fixtures that come on when the fan is switched on. The usual objection to these units is their irritating noise, which is particularly bothersome when you have no control over it, as in the case of the integrated fan-light units. We like to switch the fan separately from the light.

Higher-end ventilation units are also available in which the fan unit is mounted away from the bathroom and linked to the bath by flexible plastic ductwork. While more expensive, these fans can service several bathrooms at once, with the added advantage of being quiet. In the bathroom ceiling, the vent itself is covered by a simple, unobtrusive grille. If you are installing a spa or a whirlpool near your clothes storage area, special attention must be paid to moisture control, even to the point of installing a humidistatically controlled air-to-air heat exchanger or exhaust fan.

When I renovated my house, I made the mistake of forgoing fan units because my wife disliked the noise and the appearance. I now regret that decision, for to ventilate the baths you must open the window — an unwelcome prospect in February here in New England.

Materials

Zen states that only the spirit of a thing is important, not its outward appearance. Through the tea ceremony, and its greatest master Sen-no-Rikyu, Japanese aesthetics came to value the qualities of simplicity and rusticity, with a strong emphasis on natural materials: wood, stone, paper, and the reed mats or tatami. These two baths follow Zen principles, but in a modern interpretation.

Fixtures

There is a mind-boggling array of fixtures, faucets, and gadgets for the bath, so many that it is impossible to cover them all in this book. You can investigate the market in a number of ways. Be on the lookout for eye-catching plumbing fixtures in hotels, restaurants, and residences; peruse magazines for photos, advertisements, and articles; and visit plumbing showrooms, some of which allow you to try out faucets, showers, and even whirlpools and spas. A good re-

source is the *Kitchen and Bath Source Book* (MBC Data Distribution Publications, USA, Indianapolis, IN 46268), a phone book–size tome showing pictures and descriptions of just about every piece of kitchen and bath hardware out there. The book lists no prices, but it is updated annually and will give you a good sense of what is available and who manufactures it.

Toilets

You would think that selecting a toilet would be the simplest of all the choices you have to make when specifying a bathroom. Wrong. Scores of different models are out there; American Standard alone offers more than a dozen styles. And if the various sizes, shapes, colors, and styles are confusing, at least the names are amusing, if not downright mystifying. Manufacturers produce toilets in individual models ranging from your regular garden variety (some model names: "Plebe," "New Cadet") to toney, high-priced models drawn by famous architects and furniture designers. Most manufacturers offer collections, or "suites," of similarly styled lavatories, pedestal sinks, bathtubs, toilets, and bidets. Much of the design is excellent, with styles ranging from postmodern and minimalist, to art deco, Victorian, industrial, and what you might call 1940s apartment. Because the fixtures are offered in collections, you are able to unify your bathroom's design.

In general, high style is synonymous with high price, but some functional differences do exist between styles. The first is size: toilets come in varying widths, heights, and bowl shapes, some of which you might find more comfortable than others. Chances are, you've probably never tried a toilet on for size, but since in considering your renovation you are preparing to pay a good sum to make the room suit your tastes, you might as well get a toilet to suit your behind. The best way to choose is, well, to sit on more than one of them and see which one seems right.

Other functional differences: One-piece models are generally easier to clean; low-profile models may offer more wall space or potential for storage, such as when built into a wall unit or an extension of a vanity; many of the higher-priced toilets are quieter flushing, which may be an advantage in a master bath if you or your partner is a light sleeper. Customarily, the flush handles on toilets are on the right as you sit on the fixture. Some of the high-style toilets, though, place the handle on the left. If there is a mixture of toilets in your house, you may find yourself groping around for a handle that is not there. This may be a minor point, but in my own house all handles are on the right of the toilet except on the one in the master bath. It is incredibly annoying to have to think which room you are in just to flush the toilet.

tank above (not shown)

*one piece
push button flush*

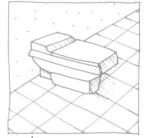
*water saver
solid plastic seat*

*low profile with siphon
vortex flushing*

telescoping style

Toilet Types

wall mount

elongated bowl

two piece with siphon jet flush

wall outlet

standard close-coupled with siphon action

In general, all the toilets manufactured by the major plumbing fixture manufacturers are well engineered and of excellent quality — you pay more for style, not for quality. Since the standard toilets are competitively priced, it makes no sense to try to economize by buying anything less than a high-quality toilet. If in doubt about which brand to buy, ask your plumber. He would know, through experience, the price-quality ratio of the various brands.

In Massachusetts and some other jurisdictions, all new toilets must be low-volume models, which flush with as little as 1½ gallons of water. While the notion of conserving water is something with which we wholeheartedly agree, some of the current generation of low-volume toilets do not always expel the waste in one flush. Repeated flushing may be necessary, which results in using as much or more water than a standard toilet. Unless you live in a jurisdiction that requires the installation of a low-volume toilet, you may want to stick with the standard models, most of which, in any case, use only 3½ gallons of water per flush. The problems with low-volume toilets will undoubtedly be solved, so it's worth asking your plumber if he knows of models that function well.

Bidets

The bidet, long a common fixture in Europe, has become *de rigueur* in the well-heeled American bathroom in the last ten years. I have to confess that before I went to Europe as a college student, I had never seen one, and the only descriptions of their use I had read were in the novels of Henry Miller. Simply put, the bidet is used to wash one's private regions; in Europe, both men and women use them. Ellen Cheever, of the National Kitchen and Bath Association, claims that regular users of the bidet consider it far superior to using only toilet paper. My wife says a bidet is great because a woman can cleanse herself without having to fully undress and bathe. It is also, she adds, an excellent place to wash lingerie or soak pots of orchids.

The bidet should be installed near the toilet so a person can use it to wash up. One straddles the device, facing the controls, which are usually placed on the wall or on the fixture's back ledge. Both hot and cold water must be available, and the controls should be easy to reach while seated. For effective cleansing, the bidet contains a spray nozzle.

Ms. Cheever says the bowl of the fixture should be long enough (an interior dimension of 20 inches is recommended) that one can reach both the front and back of one's body without shifting. Storage for towels, soaps, and accessories should be within reach when seated, and there should be a robe hook close by for clothing that's been removed.

In this bath, the dark marble countertop is softened by the radiused edge, and it plays well against the light marble backsplash, wall tiles, and crisply installed mirror. The one-piece toilet has a low profile and streamlined looks. It is matched with a bidet in a utilitarian but dignified setting.

If you plan to install a bidet, locate it conveniently and not, as we see in some speculative condo projects, wedged into a space that makes it inaccessible or unpleasant to use. Since the bidet requires a space at least 3 feet wide, this fixture must be carefully factored into the floor plan. As in all choices of fixtures and design, examine your personal habits and ask yourself if you really will use a bidet — a lot of bidets (or so it's said) have been converted to planters.

Lavatories

Sinks, like all bathroom fixtures, come in a tremendous variety of styles and materials. Probably the most common material is vitreous china, which has earned a reputation for durability, resistance to damage, easy cleaning, and reasonable cost. In the past few years, we've seen a small but growing use of painted china lavatories, which have a delicate, ornamental quality. Whether the painted decoration will stand up to daily use is something that would concern us.

Reproduction hardware looks great on this pedestal sink.

Art Deco and postmodern seem to combine in this bath with the hanging stainless-steel sink, the enamel-lined tub, and the chrome mirror with matching sidelights.

Form and materials are deftly wrought in this sculptural lavatory. The stainless-steel bowl floats in a counter of cast concrete, and the custom spigot projects from the mirror.

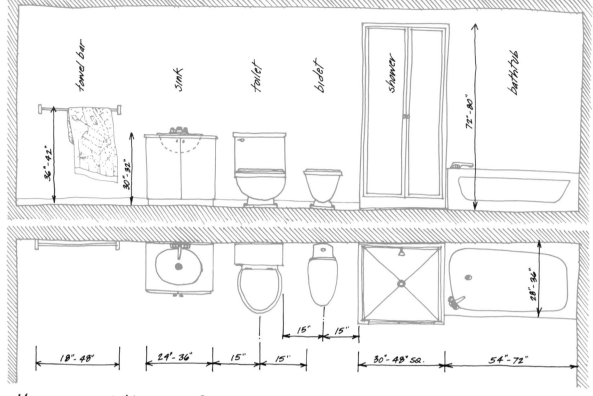

Minimum and Maximum Dimensions in plan and elevation

Another common material is acid-resistant enameled cast iron — the same material as in cast-iron tubs. These sinks are smooth, durable, good-looking, come in a variety of colors, and are competitively priced. Also available are sinks of "cultured marble," an amalgam of polyester resin, pulverized marble, and other substances. This is a problematic material, in part because it's hard for a consumer to judge the quality just by feeling or inspecting the product. "Cultured marble" is soft, and its protective layer is thin. If it wears through or is damaged by a burning cigarette, you're stuck with an undesirable appearance.

Lavatories are also available in real marble (expensive), stainless steel, and solid surfacing. Of particular interest is a solid-surfacing countertop with integral lavatories, which some manufacturers offer. Integrating the lav with the counter eliminates the rim seal around the sink, which makes wiping the counter easier and gives a much cleaner appearance. Du Pont, manufacturers of Corian, and some other manufacturers offer solid-surfacing lavs that can be bonded to a counter of matching- or contrasting-colored solid surfacing, opening up another range of design possibilities. Yet another option, and one I would explore in my next master bath rehab, is using a pair of small, stainless-steel kitchen sinks mounted under a counter of solid surfacing or stone.

Classic design and materials: china lavatory, brass and china hardware, marble counter with a decorative edge profile, recessed panel cabinet doors and drawers.

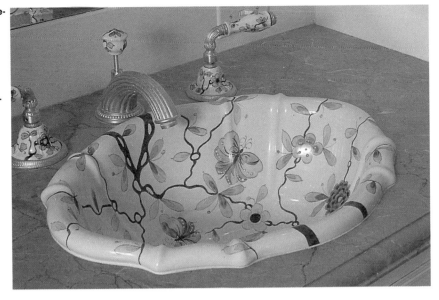

Recommended Space Requirements for Bathroom Fixtures

When thinking about the number and placement of fixtures in your new bath, keep in mind how much space is required for and between each type and the cupboards, vanities, and walls adjacent to it. The following clearances are recommended by the National Kitchen and Bath Association. These dimensions should be considered a general guide to space allocation rather than a definitive prescription for laying out the room.

Lavatories

Allocate 24 inches of countertop or wall space for each lavatory. Most of the counter space should be to the right of the bowl if the person is right handed; to the left, if left handed. Provide at least 6 inches of clearance between the edge of the lav and a side wall; 12 inches would be better. Two lavatories in the same run of counter should be at least 11 inches apart; preferably, 18 to 24 inches. For standing room, provide 21 to 30 inches from the front of the counter to the opposing fixture or wall.

Toilets

If the toilet is in its own compartment or otherwise between two walls, allocate at least 36 inches between the walls. If the toilet is between a wall and another fixture, allow 30 to 36 inches, and if between two fixtures allow 30 to 32 inches. From the front edge of the bowl allow at least 21 to 30 inches to any opposing wall or fixture, and at least 16 inches to the walkway portion of the room.

Bidets

Bidets require much the same space as the toilet. If a bidet is placed adjacent to the toilet, allow 30 to 44 inches from the centerline of one fixture to the centerline of the other. If it is placed by itself, allocate an area 36 inches wide.

Showers

Allow at least 21 to 30 inches of walkway clearance between a stall shower and any opposing wall or fixture. This space will obviously be modified to allow for the swing of the shower door — the wider the opening, the more clearance required.

Tubs

Allow at least 21 to 30 inches of walkway clearance between a tub and an opposing wall or fixture.

Towel Bars

Provide at least 24 inches of towel bar space for each household member using the bath. Bars should be far enough apart to allow easy removal and replacement of bath linens. Folded washcloths require 6 to 8 inches of hanging space (the span from the top to the bottom of the hanging cloth); folded fingertip towels, 8 to 10 inches; folded hand towels, 14 to 16 inches; folded bath towels, 22 to 24 inches; and large folded bath sheets require 36 inches of hanging space.

Exotic Fixtures

In the course of our Concord barn project, we took a field trip out to western Massachusetts to visit General Electric's Plastic House, a "concept house" built to explore the various ways in which plastics might be used in the house of the future. In the bathroom were fixtures of the "future," some of which are available today. In the shower compartment was a microprocessor-controlled set of sprayers that traveled up and down on tracks, soaking the bather with various jets of water. One then stepped out of the shower in front of a full-height body dryer, which evaporated the water on the skin with blasts of warm air. Both of these items are of Japanese manufacture and are commercially available.

Riding atop the toilet was an "intelligent toilet seat" manufactured by the Japanese firm Toto. What makes it intelligent? Well, you start by doing what you sat down on the toilet to do. Then, instead of reaching for toilet paper, you press a button on the controls attached to the side of the toilet. In response, a plastic water nozzle extends from beneath the seat and directs a jet of cleansing warm water toward your privates. The washing complete, the intelligent toilet doesn't send you away wet; it emits a stream of warm air to dry you. Depending on the particular model, you may even be able to preselect the temperature of both air and water. The seat itself may be heated, and a built-in memory stores your favorite settings so you don't have to reselect them every time you use it! The intelligent bottom washer is now being imported to the United States. Last time we checked with a Boston-area retailer, the item carried the price tag of a mere $900. So, if you're looking for that unusual gift for your next wedding anniversary . . .

If you don't think that will make a hit, you might try an "environmental chamber" such as Kohler's Masterbath Habitat. This compartment simulates nature. You can recline under sunlamps in the imagined tropics or be soothed by the gentle mists of the rain forest. You might follow this with steam. There's a sauna cycle, too, surrounding you with dry, desertlike heat. Price? Well, if you have to ask, perhaps you should take a second look at the intelligent bottom washer.

Doors and Windows

If you are lucky enough to be building a bathroom or master bedroom suite with a door leading onto a private outdoor space, you will need to install a weather-tight exterior door. Doors are available in a number of materials, including wood, metal, glass, and fiberglass. Some of them feature a core of insulating material to stop heat flow, and most have integral weatherstripping to prevent

(Left) A Santa Fe–style shower with some great details — notice the tiny lights in the beams, the well-positioned window for ventilation, the skylight, and the bench.

(Right) This skylit shower is paneled in cedar. While this species is naturally rot resistant, it may become discolored with water and soap. I would use wood in a shower only if it is coated with a boatbuilder's epoxy.

drafts. While we've had good luck with fiberglass doors, wood doors look more elegant and have a nice feel. Glass doors are made of tempered glass, which will eliminate any hazard if accidentally broken. In New England we no longer install glass doors with an aluminum frame; the aluminum conducts cold, on which moisture condenses. Instead, we use glass doors framed with wood, which look more attractive and perform better in our cold New England weather.

We feel it a false economy to forgo new windows simply to save a few dollars. Often the comfort of the bathroom can be vastly improved by installing new-technology windows, which are tighter and more energy conserving than old ones.

Windows are available in many shapes, styles, and materials. *Bay* windows are most commonly available with a center *light*, or pane of glass, parallel to the wall, and two sidelights. A *bow* window follows a gently scribed arc. Both bays and bows bring in more light while providing a nice area for plants, say, behind a whirlpool bath or soaking tub. Some of the lights in a bay or bow may be fixed, while others operate.

Casement windows are hinged outward like a door. They are operated with a crank mechanism and generally require less ef-

fort to open and close than *double-hung* windows, which slide up and down. When the window is over a counter, it's easier to crank a casement open and closed than to lean forward to move a double-hung window up or down. But because casements swing out, they are more exposed to the weather than are double-hung windows, and are thus apt to require more maintenance. Another choice is an *awning* window, which hinges at the top and opens outward, and is also operated with a crank. Awning-type windows are easy to operate and are less exposed to the weather than casements but do not admit as much air when open as either a casement or a double-hung.

If the windows are to be installed in an older house, it's important to consider how they'll complement the house's appearance; often, it's inappropriate to replace an old window with a new one of a different shape or style.

Insulated glass windows — two panes of glass bonded together with an air space in between — have been on the market for years. These may be a good, cost-effective choice for temperate climates, but for cold areas such as the Northeast we like to use the new crop of advanced windows that incorporate "low-e," or *low-emissivity*, technology. These windows (typically dual-pane) have a metallic film coating one or sometimes both panes, which helps to reflect the sun's radiant energy, keeping the heat out in the summer, yet retaining the warmth generated by the heating system in the winter. Use of low-e coatings can reduce the heat loss through a double layer of insulated glass by 50 percent.

Further efficiencies can be realized by injecting the void between the two panes with a low-energy-conducting gas such as argon, krypton, or carbon dioxide. This further improves the window's insulating value by 15 to 20 percent. A side benefit of low-e glass is that it cuts the transmission of ultraviolet light, reducing the fading of carpets, fabrics, wallpaper, and clothing stored in closets. An even more advanced technology involves suspending one or two layers of low-e plastic film between the two panes of glass, which creates a triple or quadruple glazing effect. Even greater efficiencies can be had by using an insulating material as the spacer between the panes of glass, thus further cutting the transmission of cold into the house. The most advanced windows now achieve an R-value* of 8.1, as measured in the center of the glass, or 4.6 for the window unit as a whole — a figure unheard of ten years ago. By comparison, three inches of fiberglass insulation gives an R-value of about 9.

* R-value denotes resistance to heat flow. The higher the number, the more heat is retained.

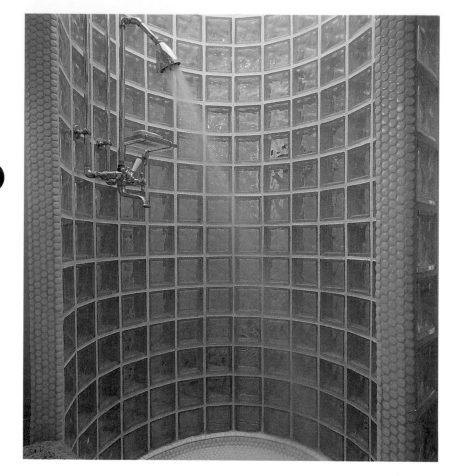

Glass block is a versatile material: waterproof, translucent, and able to be formed into straight or curved walls.

Skylights, like windows, have improved dramatically in design and performance in the last few years. In the old days, they leaked or became covered with condensation, which then dripped on walls and ceilings. They were difficult to open and close, lacked an insect screen, and came with no prefabricated flashing system — the "L" section plastic or metal material that makes a watertight joint between the roof and the skylight's curb — which made the units tricky to install. But technology has solved the condensation problem by using insulating and low-e glass, and excellent factory-built flashing kits make the new units straightforward to install and virtually leakproof. Many skylights on the market can be opened and closed without moving the insect screen, some by automatic or remote controls. You can even install sensors to open your skylights automatically in hot weather and close them when it rains.

Place your skylight carefully, considering where the sunlight will fall in the room — you wouldn't want direct sunlight to fall on the mirror, for instance. You might consider orienting your skylight to the north. Artists' studios have long been lit from above with northern light, which is diffuse and of a constant color. If you live in a hot climate, northern orientation for your skylights will keep the room cooler. Sometimes a frosted panel is used, which admits light but not the direct rays of the sun. If you're installing a skylight on a pitched or cathedral ceiling, consider having your contractor or carpenter install a horizontal soffit and a vertical sill, a detail that looks better and distributes light more widely in the room.

For natural illumination without an operating window, consider glass block. It has insulating properties, allows lots of light into the bath, and maintains privacy. Glass block would be especially useful along a shower wall.

Drywall, Blueboard, and Veneer Plaster

Decades ago, builders used to nail thin, narrow strips of wood, called lath, to the studs, which they then covered with three coats of plaster. This time-consuming operation has long been superseded by quicker techniques. Today the typical wall surface is gypsum board — large flat panels made of gypsum, a chalky substance quarried from the earth, sandwiched between reinforcing layers of paper. The gypsum board panels, typically ½ or ⅝ inch thick, are installed over the vapor barrier and secured to the studs with screws. On ceilings, we screw the gypsum board to strapping, generally 1x3-inch boards or strips of expanded metal, which we attach perpendicular to the joists on 16-inch centers. Holes are then cut in the gypsum board for windows, doors, and electrical boxes.

A number of different products composed of gypsum board are available: drywall, blueboard, fire-resistant board, and greenboard. The differences among these products are the type of paper that coats them and some additives to the gypsum.

Drywall is the most commonly used product for wall surfacing. After installation, the joints between the panels are covered with paper or fiberglass-mesh tape, which is then covered with several layers of drywall compound, or "mud." Since the compound shrinks as it dries, three coats are usually applied. When hardened, the final coat is sanded smooth. Drywall can be painted or papered and comes in a fire-resistant version for use in multifamily dwellings.

Blueboard gets its name from the blue color of its paper covering. It installs just like drywall by getting screwed to strapping or studs but is finished differently. It has a tough, waterproof paper specially formulated to bond to "veneer" plaster. The joints between panels are taped with fiberglass-mesh tape, and the whole surface is skimmed with a thin coat of plaster. The result is a clean, smooth, hard surface that closely resembles the old lath-and-plaster finish.

The wallpaper in this simple powder room is a reproduction of a 1920s design.

Veneer plastering, or skim-coating, is a job for professionals, but it has particular advantages for renovation. It easily hides any uneven meeting of old and new walls and ceilings, and because veneer plaster is a one-step process, as compared to drywall's three-step process, the plasterers are in and out of the job much faster, making way for the subcontractors who follow. Blueboard and veneer plaster is a common technique here in New England and is competitive in cost with drywall. But this is not the case in other parts of the country. Older plasterers tell us that veneer plastering is an evolution of the old lath-and-plaster techniques that were once universally practiced. In New England, plasterers never lost their skills, whereas in other parts of the country, lath and plaster was abandoned for drywall. This shortage of skilled labor may make skim-coat prohibitively expensive outside New England.

Greenboard is so named because it is coated with a water-resistant green paper. It is used as a substrate for tile on walls and shower stalls. For this application, though, we prefer cementitious backer board.

**Oak paneling and floor-
ing give a small bath a
sense of both warmth
and formality.**

Wood Paneling

Wood can give a bath tremendous warmth and character, and because wood absorbs moisture less than gypsum board, it is an excellent choice for moist environments such as the bath. On the

higher end of the cost spectrum is vertical grain fir, which, along with cedar and redwood, is both moisture resistant and beautiful. Tongue-and-groove pine would be a less-expensive choice (although, at this writing, clear pine is as expensive as cedar), as is knotty pine or pine beadboard — tongue-and-groove pine with small beading at the edge. Beadboard can be left natural for a casual, although visually busy, look or painted or stained. Beadboard is fairly cost effective in comparison to veneer plaster.

An advantage of wood is that you can stain it before it goes up, eliminating a messy job when your bathroom is nearing completion. If you plan to paint it, we recommend you prime before installation. Many people prefer merely to seal the wood and let the grain make its own design statement. Both satin and gloss polyurethane varnishes are available; use whichever one suits your tastes. Gloss is said to be more durable than satin, but for walls we think any difference would be negligible.

Keep in mind that unpainted wood will grow darker as it ages, and dark walls tend to make a room look smaller. The treatment of the wood needs to suit the overall appearance of the room. In a room with lots of things out in the open, competing for the eye's attention, wood may make the room look too busy, whereas in a room decorated in a simple, spare manner, natural wood may be just the thing.

Sensitive design and a palette of informal materials make a success out of an awkward space. Even though the angled vanity wall forces the toilet into the room, and the shape of the whirlpool niche is complicated, this bath retains a sense of harmony. The large tiles communicate "squareness" to the eye; the color and texture of the rough-troweled plaster walls mask the room's irregularities; earthen-colored grout ties the floor to the walls and ceiling, while the eye seeks the crispness of the beadboard wainscoting used as backsplash around tub and lavatory. Care in material selection extends to the wooden toilet seat, which relates to the wooden oval mirror and the tarnished look of the brass hardware.

Handmade, hand-painted tiles along with the collection of plates gives this bath a kitchen-like atmosphere.

Ceramic Tile

Tile is generally the wall and floor material of choice in commercial bathrooms. It is hard, eminently cleanable, and very durable.

There are three predominant types of ceramic tile. *White-body* tile is made from white clay, and the face glazed in any of dozens of colors. But beware: brilliant colors tend to show scratches, while more subdued colors don't. Similarly, high-gloss glazes tend to scratch easily, and those scratches will stand out on a shiny surface. Matte glazes are harder and less likely to scratch. Even if scratched, the matte finish will hide the damage quite well. A third glaze, called "crystalline" and various other trade names, looks like glass coated with sugar crystals. It is very hard and will stand up to plenty of abuse. Still another possibility is "suede" glaze, a matte finish with some shading, which is also tough and durable.

Porcelain ceramic tiles are made from a refined and very dense clay. They are fine textured, nonporous, and usually unglazed, with a smooth, silky surface that hides scratches well. Since they get their shiny surface from a polishing process rather than from a glaze, they are fairly easy to care for. Porcelain tiles are available in either a shiny or a matte surface, but not in as many colors and textures as white-body tile.

Red-body ceramic tile is made from red clay that has been refined and fired at high temperature. It is harder than white-body tile and therefore more suitable for rough duty bathrooms. Red-body tiles come both glazed and unglazed, and in a variety of colors.

In addition to manufactured tiles, we see an increasing variety of handmade tiles. Usually distinguished by their irregularities, the glaze on a handmade tile may be uneven, forming thick pools in some spots while thinning out in others. The edges of the tiles may look lumpy or there may be other imperfections, which can give your walls, floors, or counters a distinctive, handmade look. Remember, though, the more irregular the surface, the harder it will be to clean.

Also in this category are one-of-a-kind hand-painted tiles. Some tile dealers can arrange to have a craftsman paint tiles to complement the colors in your room or to repeat a pattern or motif that you specify. Hand-painted tiles command a premium price, but by buying just a few and placing them in prominent locations, you can liven up a surface made of relatively inexpensive, single-color tiles.

Some artisans will construct large-scale designs or paintings on ceramic tile. Behind the meat counter of our excellent neighborhood market, Charlie, the owner, installed at great expense and trouble a wonderful panoramic scene from his native Greece, replete with dry hills covered with olive orchards above sparkling bays dotted with the caïques of fishermen. Handmade and hand-painted tiles range in price from several dollars per square foot to twenty dollars per tile.

Red-clay quarry tile is another possibility. It is available glazed, but is most often left unglazed for a soft, rustic look. The unglazed tile will absorb oils and gradually darken, taking on the character of old leather.

Another choice is Mexican *Saltillo* terra-cotta tile. These are big tiles, 12 or 16 inches across, available in both squares and hexagons. Saltillo tiles are sun-baked rather than fired in a kiln, which gives them a wonderfully funky look. We used Saltillo tiles on the first floor of the Concord barn project to stunning effect. While Saltillo is mostly used on floors, there is no real reason it could not be used on walls. These tiles are available unglazed, glazed, and sealed with an epoxy sealer. Like other types of terra cotta, Saltillo will stain if left unsealed, but even when sealed it will most likely

This farmhouse bath is unabashedly country — rough walls and exposed stone make fitting surfaces for a rustic bath.

stain and darken with age. Yet if there is one material that actually improves with age and blemishes, it is Saltillo. If it gets intolerably marred, Saltillo can even be sanded down and resealed.

Recently, Italian tilemakers have begun producing glazed ceramic tiles that have the look and warmth of terra cotta but are stain resistant and easily washed with a sponge mop.

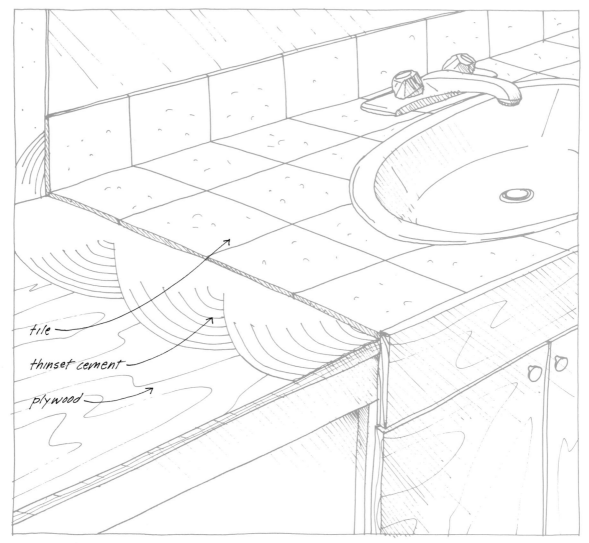

tile —

thinset cement —

plywood —

Tile Installation at Bathroom Counter

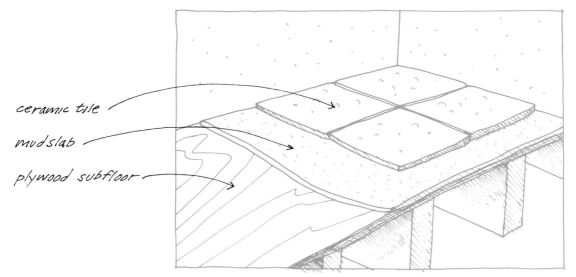

ceramic tile —

mudslab —

plywood subfloor —

Mortar Bed Method at Bathroom Floor

Jade green tiles make up this combination bath and shower niche. The dividing wall is open, to preserve views for both shower-takers and bathers. The walls are salvaged maple flooring.

Size of Tiles

Manufactured tiles are available in many sizes, the smallest being 1-inch square. These are linked into sheets with paper or strands of silicone. A 1x2-foot sheet makes it possible to install 288 1-inch tiles in a single operation, saving a tremendous amount of time. Selecting the size that is right for your bath is partly a matter of scale, partly a maintenance issue. Little tiles can make a small room seem larger, but the bigger the tile, the fewer the grout lines to keep clean. Also, larger tiles can help mask irregularities in an oddly shaped room.

Installation of Ceramic Tile

There are several methods of setting tile. Installation on walls is not as demanding as on the floor or countertop. In a dry location, the tiles can be glued directly to drywall with *mastic*, a gooey, organic adhesive that is applied with a notched trowel. In wet locations, you should use the *thin-set mortar* method, which we will discuss shortly.

On floors, the traditional *mortar-bed method*, or "mud job," is still frequently used in the western United States. The installer lays a bed of portland cement 1 to 2 inches thick on a base of exterior-grade plywood, covered with building paper and chicken wire. When properly executed, a mud job can last for decades. Most failures of a tile floor, such as breakage of individual tiles or cracks in the grout, stem from instability of the substrate rather than from flaws in the materials. The mortar-bed method is extremely stable — the tile setter has, in essence, built a rigid masonry slab — so there is no cracking of the grout or tile. Usually, a mud job is more expensive than alternative methods.

In most of the United States, tile is installed with the *thin-set method,* so called because the mortar can be as thin as 3/32 of an inch. On floors, the installer must lay a deck of exterior-grade plywood, secured to the subfloor with building adhesive and screws so it will not move. In the shower, greenboard can be used, but we prefer *cementitious backer board,* a cement and vermiculite tile substrate available at the lumberyard. The installer then spreads mortar on the backer board with a notched trowel and presses the tile into the mortar. Several different kinds of mortar can be used, including a high-strength epoxy.

A variation of the thin-set method involves the application of an organic mastic adhesive over exterior-grade plywood. This is quick and inexpensive, and while it is fine for walls and ceilings, we do not recommend it for wet locations such as showers and countertops. If water penetrates to the plywood substrate, it can delaminate the plywood or loosen the tiles. Since plywood expands and contracts at a different rate than does tile, it is nearly impossible to maintain tight grout joints between the backsplash and the counter and at the sink's rim seal.

Grouts

There are a number of grouts on the market, the best being *epoxy grout.* It is strong yet flexible enough not to crack with the expansion and contraction of substrate materials. It is extremely stain resistant, and its colors, even white, will not darken over time. Unlike traditional grouts, epoxy contains no cement at all. Instead, it is composed of epoxy resins, plasticizers for flexibility, and colored filler powder. It demands special application techniques. Each batch has a working time of about 45 minutes, during which it must be applied, or "packed," and then cleaned. Once the chemical reaction that hardens the grout occurs, the grout will be permanently stuck wherever it is. Epoxy grouts are about three times the price of other types; according to John Pascoe, they are the product of choice.

In the many years before the advent of epoxy grouts, tile setters used *commercial portland cement grout,* which is simply a mixture of 10 percent commercial portland cement and 60 percent sand.

There is nothing wrong with this type of grout; it is dependable, economical, and its properties and application techniques are well known and understood. Commercial portland cement grout has a long working life but must be "wet-cured" in order to harden. After the grout has been applied and cleaned from the joints, the tile surface is sprayed with water and covered with a sheet of polyethylene. It takes twenty-four to forty-eight hours to cure.

Commercial portland cement grout is inexpensive, and if properly installed performs excellently, although it will stain and darken over time. It can be used with any kind of tile but is ideal for applications where the grout joints are wide.

There are two additives, or "admixtures," commonly combined with commercial portland cement grout to alter its properties. Addition of 5 percent methyl cellulose (wood pulp) to the standard brew will produce *dry-set grout*, a formulation especially used with terra cotta and other porous tiles. The wood pulp helps the grout to retain water that would otherwise be sucked out by the tile. If used on less porous tiles, dry-set eliminates the need to keep the tile hydrated with the plastic sheet.

The other admixture is latex, which can be added to both dry-set and to the standard portland mix. It improves the grout's bonding both to the tile and to other substrate, its compressive and tensile strength, and the grout's stain resistance. But John Pascoe warns that latex can leach out into some types of tile, ruining its appearance. Latex comes as a liquid or a powder. To prevent leaching, it is imperative that the latex be mixed strictly according to the manufacturer's instructions.

There are two other types of grout: a silicone-based mastic grout, which is something like silicone caulk and is still available in hardware stores and home centers. This type of grout is not widely used commercially. Also, a new commercial grout has just come on the market, called cementitious mastic grout. It is extremely flexible and is applied right out of the container.

Counters

High-Pressure Laminates

Of all the counter materials on the market, the most widely used is plastic laminate. Many people know it only by the brand name Formica rather than by its generic moniker. There are several brand names for this smooth, easy-to-clean surfacing material — Formica, Wilsonart, and Micarta — but the product is basically the same: a series of thin layers bonded together under heat and pressure.

The topmost layer is melamine, a clear plastic coating that protects the lower layers from abrasion and water penetration. Next

This tightly composed vanity explores subtle conversations between materials: the granite counter speaks to the stone-pattern ceramic tiles in the language of shape and texture; the tiles to the mirror, via rectangularity; the mirror to the lighting fixtures, via smooth texture and chrome edge. The round-ness of the fixtures draws the eye back to the smooth round sink, which completes the con-versation and ties the composition together.

is a layer of pigmented or decorative paper for color and pattern, and beneath that are several more layers of craft paper impreg-nated with phenolic resin, to give the laminate body and impact re-sistance. The finished material is less than 1/16 inch thick.

Countertop Construction

To make a solid countertop, the laminate must be bonded (with con-tact cement) to a substrate of particleboard. Particleboard is pre-ferred over plywood (with the added bonus of being slightly less expensive), as it is harder, offers better impact resistance, and is smoother, which allows a better glue bond with the laminate. Parti-cleboard has no grain structure that could be telegraphed through to the finished countertop. Also, particleboard has a coefficient of expansion similar to laminate, which means that the two materials expand and contract at the same rate, as the ambient temperature and humidity changes. Plywood's coefficient of expansion is higher, which can weaken the glue bond and cause seams to open up.

The standard countertop thickness is 1½ inches, built up of two layers of ¾-inch particleboard. The top layer expresses the full shape of the countertop; beneath it, smaller strips, or *fillets,* are glued and pneumatically stapled to the edge and all points where the countertop will contact the base cabinets. The finished counter-top is then screwed to the base cabinets from underneath.

Edge Detail

The most common edge treatment is "self edge," meaning that a strip of the same laminate used on the surface is attached to the front edge of the countertop. Where the horizontal and vertical surfaces meet will be a small dark line where the laminate's core is exposed. Self-edging is usually the least expensive option, and visually the most elegant, in that the edge will not draw attention to itself.

If you like the simplicity of a self edge but object to the dark line, consider using *color-through laminate,* in which the color is consistent throughout. It is beautiful (and expensive) and can be used to produce the illusion of a solid countertop. Laminate manu-facturers produce color-though laminates in the same colors as reg-ular laminate, so you can mix the two types without worry of color differences.

Another approach is a *decorative edge detail,* the most common being to fasten a hardwood strip around the perimeter of the counter. Oak, birch, and maple are the most commonly used, either left natural or stained. The laminate abuts the hardwood at the top edge of the counter, so you don't see the brown interior of the laminate as with self-edge. Also, hardwood will absorb the nicks and blows the edge is sure to receive, whereas a self-edge might chip.

The edges of the wood itself can be profiled in several ways. They can be left square, "eased" slightly, or they can be quarter-rounded, half-rounded, or profiled with a more elaborate ogee.

At "This Old House," one of Norm's favorite edge profiles is a simple 45-degree *bevel,* or *chamfer,* to ease the transition be-tween horizontal counter and vertical edge. Norm runs the laminate over the hardwood edging, then chamfers both wood and laminate with a chamfering bit in a router. The brown line of the laminate is less noticeable next to the wood, and the joint between wood and laminate less vulnerable to penetration by dirt and moisture. You can accent the chamfer by laminating a strip of plastic, contrasting-color wood, or even aluminum or brass behind the outermost layer of wood. You can also build up a decorative edge without any wood by using layers of plastic or metal.

Prefabricated edge treatments come in a variety of colors and wood grain finishes. They install by snapping into a groove cut into the front edge of the counter.

Since edge treatments can be installed in different ways, ask your fabricator if there will be a noticeable seam between the edge treatment and the laminate and if the seam will be impervious to water, which will destroy a counter if it penetrates to the substrate.

It is also possible to put decorative *inlays* of wood, metal, or a contrasting color of laminate into the surface of the countertop itself. But because this entails exposed joints in the laminate, it increases the chance of water penetration and eventual failure. If you want embellishments in the horizontal surface, it is best to use a solid surfacing material, discussed later in the chapter.

The Backsplash

Standard backsplash dimensions are 4 inches high by ¾ inch thick, and the fabricator of your laminate countertops will probably supply them in the same color laminate as your counters unless you specify otherwise.

The backsplash is bonded directly to the wall with construction adhesive, and the top edge is caulked with silicone or latex caulk. There are two methods for making the transition between counter and backsplash. One is to install an aluminum cove molding at this juncture; the other is to caulk it with a thin bead of silicone caulk.

Another approach to finishing off the wall edge of the countertop is to bond, directly to the wall, a sheet of laminate that covers the entire wall space between the top of the counter and the bottom of the mirror or cabinet, eliminating the need for a backsplash altogether. Typically, this is done prior to installation of both mirror and counter.

Postformed Countertops

One way to integrate the backsplash, countertop, and edge detail in a single unit is to install a postformed countertop. *Postforming* refers to the process whereby laminate is heated to make it pliable and then bonded to a formed substrate. Since the laminate cannot be bent to extreme radii, the backsplash melds into the countertop in a sweeping curve rather than a sharp edge, and the front edge is given a semicircular wrap or a blunt bullnose. Some postformed counters feature a slight rise at the edge of the horizontal surface, which keeps water from dribbling into the vanity or onto the floor.

Postformed counters are available in a limited range of colors and patterns, but additional nonstock colors may be special ordered. The laminate used on these counters must be slightly thinner than usual (the industry standard is .042 inch as opposed to .050 inch for conventionally formed counters and .030 inch for cabinets and doors). The chief disadvantage of these counters, aside from their rather old-fashioned look, is that in an L- or U-shaped bathroom the joints between runs of counter will have to be made like

85

Another tightly composed vanity, reminiscent of Mies van der Rohe, combines the warmth of ash with the taut coolness of a solid-surfacing counter, hinged mirrors divided by vertical fluorescent lights, and aluminum windows.

picture frames, in 45-degree miters. Making these cuts in the field to close tolerances is difficult. Another problem is finishing the end of the countertop. In a self-edge detail, the top laminate covers the vertical laminate of the counter edge. Since this is tough to do with post-formed counters, a wood or laminate endcap must be fabricated and fitted to the counter end.

Seams

The Achilles' heel of a laminate countertop is the seams. Water penetration to the substrate will cause the glue bond between laminate and substrate to fail, the plywood to rot, and particleboard to expand (as much as 75 percent). In any event, the counter will self-destruct. Laminate comes in 5x12-foot sheets, which means that in most bathrooms there should be no need to have a seam in the countertop.

Considerations When Choosing Laminates

High-gloss laminates scratch very easily and show scratches readily. Since laminates cannot be sanded or treated to remove scratches, we avoid glossy laminates for counter duty. A matte finish is best for scratch resistance.

Dark, solid colors show both blemishes and dirt and we generally advise our "This Old House" homeowners against using them. Laminates are also available in a variety of textures — leather, slate, stone, and an embossed grid.

Another thing to consider with laminates, as with all finishes and surfaces in your new bathroom, is how enduring will be the color and style from a decorating standpoint. A bright purple floral pattern might be the rage this year, but how will you feel about it three years down the road? We generally advise our homeowners to choose a neutral white or gray, for these colors will accommodate many changes in the paint and wallpaper scheme.

Solid Surfacing

Solid surfacing is an acrylic-based plastic invented by the Du Pont Corporation and is still widely known by Du Pont's trade name Corian. It was originally available only in white and off-white. Now there are similar products by other manufacturers — Avonite, Fountainhead, Surell, and Gibraltar — in over fifty patterns and colors, including some beautiful granitelike patterns.

Solid surfacing is considered by many, including most of us at "This Old House," to be the best choice for bathroom counters, and is increasingly being used in place of ceramic tile for shower stalls, walls, and tub surrounds. Solid surfacing is a structural material that can be used to make cantilevered overhangs along edges of counters, and it is dimensionally stable so it will not warp or shrink. The color and pattern are homogeneous throughout, so the material can wear for many years. It is resistant to water, soap, heat, scratching, denting, and staining. If cut or damaged, solid surfacing can be sanded smooth again with an ordinary orbital sander or by hand. In the worst case, portions of the counter can be filled with a filler that matches the color of the material or the bad spots can even be cut out and patched. Solid surfacing is easily cleaned with standard household cleansers.

All these wonderful attributes are not without cost, however. Solid-surfacing countertops are three to five times as expensive as high-pressure laminate, but less costly than granite or marble. For tub and shower surrounds, the cost of solid surfacing is probably equal to or slightly greater than tile.

Although solid surfacing can be worked with ordinary woodworking tools, it is not a project for the average homeowner. In fact, most contractors and carpenters subcontract solid-surfacing work to a specialist. Specialists, in turn, tend to use just one or two brands of material, since each has different material properties, working characteristics, factory-recommended fabrication techniques, and proprietary seam adhesives. Beware of the fabricator who claims he can work with equal facility in all brands. The fact that each manufacturer conducts special training seminars and publishes fabrication handbooks and regular product bulletins, and the fact that some fabricators give their personnel six months training before turning them loose unsupervised are strong arguments for going to a shop that specializes in just one or two brands.

Detailing of solid surfacing is limited only by your imagination — and, of course, your budget. A variety of edge details can be applied, from a simple quarter- or half-round to chamfers and more complex ogees and laminated "feature strips." It is a relatively easy matter to build up the edge thickness to 2 or 3 inches to give the countertop the look of a massive piece of granite or stone. Inlaid patterns in the counter surface or edge detailing can be made without fear of damage from water penetration. In general, the more complicated the detailing, the more costly.

Solid surfacing comes in ¼-, ½-, and ¾-inch thicknesses, and in various size sheets. The thicker sheets can be cantilevered to form overhangs at the edges of counters. The usual rule is that ½-inch-thick material will tolerate a 6-inch overhang while ¾-inch-thick stock will tolerate a 12-inch cantilever, although some fabricators will extend only to 9 inches. Some brands, such as Corian, do not require a particleboard or plywood substrate; others, such as Avonite, do.

Many fabricators, designers, and architects consider Corian to be the standard by which all others are judged. It has been in production for more than two decades and has a track record of dimensional and structural stability and resistance to environmental factors such as strong sunlight. At this writing, however, Ralph Wilson Plastics has just weighed into the field of solid surfacing with its offering, Gibraltar. Colors and patterns are keyed to those of its high-pressure laminate, which means that to save some money you could use solid surfacing for the countertop and high pressure laminate for the backsplash, or laminate for the countertop and solid surfacing for the edge detail. Early reports on Gibraltar's fabrication qualities are favorable.

There are several factors to consider when choosing among brands. First, of course, is the color and pattern of the material — Avonite, for instance, is still favored by designers for its colors. Another factor is the stock dimensions. If you are planning to panel a shower that is 36 inches wide, it would make sense to buy the material in that width to eliminate a seam. This consideration would knock Corian out of the running, as it is available only in 30-inch-wide sheets.

A third factor is the accessories, or "shaped products," offered by the manufacturer. Du Pont, for instance, manufactures thirteen sizes of sinks and lavatory bowls that your fabricator can bond right into your countertop, an option that, in and of itself, in our opinion, is almost worth the price of solid surfacing as it eliminates one of the most vulnerable areas of the bathroom counter, the rim seal.

Marble is one of the warmer stones, which can give a room a sense of rich and controlled power — like a purring jaguar. Here, two baths show off its elegant qualities.

A fourth, and critical, factor in your decision among brands should be the availability of a top-flight fabricator certified to work the material of your choice. Even though Du Pont, for instance, guarantees Corian against manufacturing defects for ten years, failure to fabricate it using certified techniques and materials will void the warranty. Du Pont certifies fabricators through a network of local training seminars. (This policy puts the home craftsman in a bit of a bind, since doing your own fabrication may hinder your ability to make later warranty claims.) Because the use of solid surfacing may represent a substantial percentage of your bath renovation budget, evaluate your prospective fabricator carefully. Ask for a straightforward explanation of his or her and the manufacturer's warranty policy and for the names of several customers.

Stone

Many people consider stone to be the ultimate material, not just for counters but for floors, walls, and shower stalls. Stone is beautiful, and so durable as to be permanent. It is hard and conducts heat efficiently, which means that in any climate other than a tropical one, the installation of some form of radiant heat behind the stone is in order. It is also among the most expensive of materials, depending on the type of stone and the area of the country you live in.

Installing stone is much like installing tile. On vertical surfaces, it should be applied to cementitious backer board. On floors, it must be laid on a solid, nonflexing base. If you plan to use large slabs or heavy tiles, make certain the structure of your floor is sufficient to carry the additional weight. Also, keep in mind that very thick tiles or stone slabs may cause the finished elevation of the bathroom floor to be uncomfortably higher than adjoining rooms.

On most of the East Coast, slate is common and, therefore, fairly cost effective. It is readily available in sheets of various thicknesses. Slate is soft and easily workable, which lends itself to advanced do-it-yourselfers. It cuts with a diamond-tipped circular saw blade and can be smoothed with a belt sander. Like all stone, it can be glued to steel, wood, or just about any other material with epoxy. Slate is available in two surface textures: polished, which has a silky, soapstone-like feel, or with a natural *cleft*. Slate's chief disadvantage is its color, black, which makes any dirt, soap stains, or water drops stick out like a sore thumb. Unless you are very meticulous, your slate bathroom will always look dirty.

For high-end master bathrooms, marble is the material of choice for floors, walls, and shower enclosures. It is beautiful, rich and warm in appearance, and is available in a variety of colors and textures. Marble is porous, which makes it easily stained, and it should, therefore, be sealed with masonry sealer. To increase stain resistance, marble can be polished to a high gloss.

A *honed* finish will stand up better to rough duty. The satiny texture can be scoured with steel wool to remove some stains. Those that remain will be hidden somewhat by the stone's nonreflective finish. You may have to be persistent to obtain honed marble. Most of the imported slabs come prepolished, and domestically produced stone is polished before it leaves the yard.

While solid slabs of marble are expensive, Mexico has recently begun exporting quite acceptable grades of inexpensive marble tiles, so check twice before you reject this material on the basis of price alone.

Granite is a very elegant stone not commonly used in bathrooms, but, as with all stones, we see no reason it could not be used. It is about 30 percent more expensive than marble, although this comparison is only approximate because of the many grades of granite and marble to choose from. Rare grades of both stones will be more costly then plentiful grades, of course.

Granite has a tighter grain structure than marble, so it looks very uniform or monolithic. Slabs for countertops are usually polished to a smooth, high-gloss surface that can be washed with soap and water and touched up with window cleaner. Polished granite is impervious to stains. Like slate, black granite shows every smudge and soap spot, whereas a more variegated pattern does not. Honed surfaces in granite tend to be dull and lifeless as well as being more vulnerable to stains. As with marble, granite is available in tiles.

Frank Lloyd Wright was one of the early experimenters in the plastic qualities of cast concrete. Architects and designers are still attracted to it for its ability to be formed, colored, and textured. The cast-concrete tub here gives way to a shower room and outdoor pool.

Buying stone in slab form involves a much different procedure from buying other materials for your bath. Stone dealers, kitchen and bath stores, and designers usually have small sample pieces to show what is available, but placing your order solely on the basis of the sample is risky, especially with marble that has a prominent vein structure, or *figure,* the full extent of which will not be revealed in a small sample. Stone is a product of nature, and the color and figure will vary even through the same quarry.

If you are ordering your stone from a local stoneyard or distributor, go there with your mason, fabricator, or installer and select the stone for your job in person. If the distributor is too far away, ask for a sample from the lot from which they will fill your order. For our adobe rehab in Santa Fe, New Mexico, our homeowners selected a pink Mexican marble for their kitchen and bath countertops. While they made their initial color selection from architectural samples, they traveled to the fabricator's yard in Juarez, Mexico, to make their final selection.

In the course of your visit to the stoneyard, you may find that the dealer is willing to offer more attractive prices on stone he's already got in stock. A way to get stone at an even better price is to

A stone grotto retreat — lacking only naiads.

call a large stone fabricator and inquire about remnant slabs. A yard that turns out the facing of office buildings and other large projects inevitably ends up with leftover pieces, some of which may be big enough for your bath. If you are willing to do a little legwork rather than simply ordering through a dealer or stonemason, you may find a real bargain.

●

To cut the stone to the dimensions of counters, a full-sized pattern or template is created. The template is made of a waterproof material such as thin plastic, which will stand up to the large quantities of water required by the diamond-tipped wet saws used in stone cutting. Most slabs used for bathroom counters are ¾-inch thick, although thicker slabs are available. The length and width of slabs varies from piece to piece, but because of the weight and brittleness of stone, 8 or 9 feet long, and 4 or 5 feet wide is about the maximum-size slabs quarries like to handle — large enough, certainly, for the biggest bathroom counter.

Sometimes the slab simply sits on top of the base cabinets and is secured by silicone or epoxy. In other installations, it rests on a plywood substrate.

The exposed edges of the stone should be slightly rounded, or "eased," since a sharp edge is likely to get chipped or cut some unfortunate child's forehead. The edge can also receive a quarter-round or a more decorative profile.

Most stone floors will be laid in the form of tiles, which are purchased and installed like ceramic tile. However, in our Santa Fe, New Mexico, project, we used flagstone which was shipped to the job site in irregular slabs. The pieces, which varied from one to three feet square, were sculptured to fit together as each was laid.

Here in the Boston area there is a particularly beautiful variety of pink granite known as Milford granite (after the town in which it is quarried). Slabs of Milford can be huge — up to 6 feet square — and the possibilities of using a single slab of it for a bathroom floor has intrigued me since I first saw the stone. Obviously, such a massive piece of rock would have to be designed into the project from the beginning and perhaps even installed before framing the walls. Clearly, smaller slabs of Milford would be easier to install.

If you go to the quarry to select and carry it, Milford is surprisingly cheap — competitive with tile. Similarly, locally quarried bluestone is quite inexpensive and also makes an unusual and beautiful floor. Perhaps there are inexpensive sources of local stone in your area. With radiant floor heat and the availability of good stone sealers, don't limit your imagination to the old standards in flooring materials.

A modernistic interpretation of the old-fashioned washstand. A stainless-steel bowl and high-tech Swedish faucet are visually linked to the black toilet and tub by the sculptured stainless-steel shelf.

Sealers

Unglazed tile and stone should be sealed to prevent water penetration and to increase stain resistance. A mixture of equal parts of linseed oil and mineral spirits does a good job, but requires several thin coats. If applied too heavily and allowed to stand on the surface of the tile, it will form a gummy residue. Floors treated with this concoction should be resealed whenever it appears the coating is breaking down.

Epoxy sealers avoid the drawbacks of the linseed oil–mineral spirits mix. They are typically applied once, penetrate into the tile or stone, and dry to a tough, durable finish. However, epoxy sealers may darken the color of the floor unevenly or give it an "artificial" look.

Silicone sealants, similar to the compounds used to waterproof shoes and gloves, may also deepen the color of tile and stone, but in our experience tend to do so evenly. These sealers more closely mimic the rich tones imparted to the floor by the linseed oil–mineral spirits mixture.

Oil-based sealers such as Val-Oil do a good job sealing the floor and also deepen the color of the masonry, but do so evenly. Before you select a sealer for your tile or stone — or allow your mason or tile setter to apply one — we recommend you test a variety of products on samples of your floor. Each product imparts a different color quality to the floor, and if you've gone to all the expense and trouble to lay tile or stone, you will want it to look just right.

•

Although the materials mentioned above represent the most common of those used in bathrooms, the list is by no means exhaustive. Here are a few more possibilities:

Stainless steel is tough enough to withstand constant use, stands up to high temperatures, will not stain or rust, and is easy to clean and sanitize. For all these reasons it is used in virtually every commercial kitchen in the country. We see no reason it could not be used in a bathroom as well. Some stainless-steel lavatories are available commercially, but any unique application of the material to a bathroom would have to be custom designed and fabricated. While the cost of stainless steel is high and its "warmth factor" low, its cool postmodern look makes it an intriguing possibility in certain houses and bathrooms.

Copper is another material used in kitchens that could be adapted to the bath for counters or even a shower surround. Copper is expensive, of course, but its deeply burnished warm glow is gorgeous. Copper is soft, so it will dent easily. To keep it from tarnishing, it must be polished or coated with a lacquer finish, which itself is soft and difficult to maintain.

Contemporary architects from Frank Lloyd Wright to I. M. Pei have been taken with the structural and aesthetic properties of concrete, and have used it in original ways in their work. Concrete's advantages in a bathroom are obvious — it is totally waterproof and virtually indestructible. Because it is difficult and expensive to form and finish to a high standard, however, it is out of reach for most bathrooms. Yet the material has stunning visual properties, and in the right bath and with the right budget, the possibilities are manifold.

In some restaurants and residences, we've seen lightweight concrete counters, dyed to various colors such as rose, gray, and black. This material can be cast in any shape and thickness. Cast concrete would be suitable for floors and shower surrounds as well as tub pedestals.

Most woods are not suitable for counters, showers, and tub surrounds, but there are several exceptions. Teak has long been the premier boatbuilding wood because it is dimensionally stable and resistant to rot and marine borers. Teak is expensive, but its rich, mocha-colored grain finishes beautifully and would lend real distinction to a bathroom vanity.

Floors

Aside from the one in the kitchen, no other floor gets as much wear and abuse as that in the bathroom. And yet, of all the design elements — cabinets, counters, wall coverings, and paint colors — it perhaps most strongly influences the room's overall look. Ceramic tile will give the room a crisp, functional look; marble, a formal look; terra-cotta tile, a French country flair; and vinyl, a . . . well, a standard bathroom look.

When you choose flooring, take into consideration the type of bathroom you're renovating and the duty it will see. Using cost-effective vinyl in an elaborate master bathroom is as pointless as using an expensive flooring material in a utility bath. In the following chapters we discuss what types of flooring are appropriate, but let's review the options here.

Vinyl Flooring

Sheet vinyl is by far the most common choice in bathroom flooring. Manufacturers call it resilient flooring because it provides a slight cushioning that is easier on feet and legs than an unyielding material like ceramic tile. Vinyl is warmer than most other floor materials, is impervious to water, resists stains, has fair traction, and is cost effective.

Most varieties of sheet vinyl come in 6- or 12-foot widths. The ideal is to cover the whole floor with a seamless sheet, for a visible seam will detract from the floor's appearance, and, more important, the seam is the most vulnerable part of a vinyl floor.

Most vinyl flooring is composed of three layers: bottom backing, usually made of vinyl or felt, the middle design layer, containing the floor's color and visual pattern, and the top wear layer, which protects the color and pattern from being rubbed away by traffic. Manufacturers now apply a vinyl wear layer, which eliminates the need to wax the floor. This type of sheet vinyl is promoted as "no-wax."

No-wax vinyl floors do require care. A spokesperson for Armstrong World Industries, one of the major vinyl manufacturers, said that all vinyl floors lose their shine soon after they're installed. If you don't mind a dull-looking floor — and many people don't — you needn't do anything except clean as necessary. Wet mop the

Integration of custom teak counter with support post as divider — a frosted Plexiglas panel is a permanent shower curtain.

floor to remove dirt particles that would otherwise grind in and dull the surface. Some manufacturers advise customers to reserve one bucket and mop for cleaning and a second bucket and mop for rinsing, because the film left from inadequate rinsing can make the surface dull.

In addition to no-wax floors, one manufacturer, Mannington Mills, makes an extensive line of what it calls *"never-wax" vinyl* flooring. According to Mannington, its wear layer is much thicker and, therefore, longer-lasting than that of a standard no-wax vinyl floor.

Armstrong counters that the difference in the thickness of the wear layer has no practical consequences, and that while its wear layer is thinner, it will never be "walked off" in normal household use. If you want a tougher finish, Armstrong recommends you buy its line with a vinyl surface coated with a urethane layer, which keeps its shine longer (but still needs periodic polishing). Congoleum and Tarkett, also major manufacturers, offer similar products.

Since all vinyl flooring needs maintenance, one of the factors you might consider when selecting a brand is what type of maintenance you're willing to do. Also pay attention to the thickness of the overall flooring, not just the wear layer. Generally, the thicker the vinyl, the more durable (and expensive) the floor will be. You would be well advised to go to a retailer or installer who handles several manufacturers' products and get him to explain the advantages and disadvantages of each, as well as recommend cleaning agents or polishes. Some manufacturers sell cleaning products formulated especially for their brands.

Some vinyl flooring is embossed with indentations that enhance the design or create a pattern of light and shadow that makes blemishes less noticeable. Note that light colors tend to hide scratches and blemishes, while dark colors make them stand out. Embossing may make the vinyl harder to clean. Also keep in mind that vinyl can fade after prolonged exposure to intense sun, just as draperies and upholstery do. This might not be a danger in most bathrooms, but it's worth taking into account if the flooring is to be installed in a sunny space or next to a glass door.

Another vinyl product, called *inlaid flooring*, gets its pattern and color from a different process. Instead of a printed design layer, thousands of colored vinyl granules are fused onto the flooring by heat and pressure. Armstrong claims this process results in greater depth and richness of color (and, not surprisingly, greater cost).

Vinyl flooring is usually installed over a plywood underlayment that is first nailed to the existing floor to even out any irregularities. However, one vinyl product can be laid directly on any existing floor. Called *interflex flooring*, it contains a vinyl backing that can span minor irregularities. Interflex is generally chosen to save on installation costs.

Linoleum

Years ago, linoleum was a favorite flooring material. It was made of linseed oil, wood dust, cork, and resins. Its drawback was that it required waxing, and after no-wax vinyl flooring appeared on the market, demand for linoleum plummeted. Armstrong, the last major American maker of linoleum, stopped production in 1974. But fashions go through cycles, and now there is a small but growing demand for the look of linoleum — especially for a patterned field with a contrasting border. Linoelum can still be obtained through a few companies that import it to the United States. Recently, vinyl imitations of linoleum patterns have been introduced, and it's possible to order flooring with one pattern for the field and another for the border. If you're willing to pay premium prices, you can find other kinds of vinyl flooring, most of them tough, durable materials used in commercial buildings.

Installing Vinyl Flooring

The critical step in installing vinyl flooring is to prepare the subsurface. What's needed is a solid, uniform base, free from irregularities. Sheet vinyl can be laid over an existing floor (vinyl or another material) as long as it is smooth and stable. Any ridges must be sanded down and filled or they will cause visible high spots and be felt underfoot. A leveling compound similar to spackle can even out some depressions, and sanding and scraping can bring down some high spots. However, the installer may be unwilling to guarantee the floor will remain smooth, since it could still be pushed out of shape by pressure from the old flooring underneath. The safest course is to rip out old vinyl or linoleum and start anew by putting down plywood underlayment. Removal is a more critical issue if the old flooring contains asbestos, as many old linoleum tiles did. There appears to be no health hazard in leaving the tiles on the floor and covering them. The hazard, to the extent one exists, is in breaking the tiles and scattering asbestos fibers into the air. In making plans for floor removal, make sure the installer will discard the material in accordance with state or local laws and codes.

Underlayment is important even in a new house, for the wooden subfloor constructed by the home builder is usually too rough and uneven a base for vinyl. While lauan plywood (a species of mahogany) is the most common underlayment, the vinyl-flooring specialist for "This Old House," Paul Vogan of Belmont Flooring in Belmont, Massachusetts, prefers Structurwood Underlayment, made by the Weyerhauser Company.

In Japanese aesthetics, an object is valued for its age and rusticity. The old character of the doors, floor, and tub, along with the Shaker-style stool and towel rack, gives this bath an oriental calm. The old plank floor has been stained and sealed to help it stand up to the inundations of dripping bathers.

The importance of underlayment is emphasized by the fact that vinyl manufacturers will not warrant their products if installed without it. Some installers *will* guarantee their installation for at least a year, although few, it seems, are willing to put this in writing. The haziness of some guarantees points up the importance of discussing warranty coverage before agreeing to use a particular installer or product.

Installing vinyl flooring is slightly more complicated if the bathroom is built on a concrete slab over an unventilated crawl space. Moisture can penetrate the floor from below and, trapped by the vinyl, form bubbles. If the builder has installed a vapor barrier over the soil before pouring the concrete, there should be no moisture problem. If the bathroom sits above a crawl space, moisture problems can be alleviated by installing a vapor barrier over the ground in the crawl space.

Vogan recommends a simple do-it-yourself test to determine whether moisture will be a problem in a slab floor. Tape several pieces of plastic to different parts of the floor. Check the plastic every day to see if it has trapped any moisture. If, by the end of the week, no moisture buildup is noticeable, the floor has probably been sealed against vapor transmission. Also, check for depressions or unevenness before installing the vinyl. A fine cement can be applied to level the concrete.

Vinyl Tiles

For low cost and ease of installation, it's hard to beat vinyl tiles. They come in 9- or 12-inch squares, and have long been favorites of do-it-yourselfers. Vinyl tiles are particularly suitable if the bathroom is built on a concrete slab. They're less suitable for use over a wooden floor, since wood may shift and cause tiles to come loose. Vinyl composition tiles have now replaced the older squares made of vinyl and asbestos. Generally tiles cost less than sheet vinyl.

Preparation for vinyl tile installation is the same as for sheet vinyl. Easiest to install are the tiles with an adhesive backing exposed by peeling off a layer of paper; but we have more faith in tiles that are secured with a separate application of trowel-applied adhesive.

Some floor tiles have a no-wax finish and are constructed like sheet vinyl; others are solid vinyl and are more expensive. Some solid vinyl tiles imitate the appearance of wood, ceramic tile, or slate. Tiles are available in commercial grade also.

Yet another product is cork tiles, which are covered by a wear layer of clear vinyl. These combine the look and springiness of cork with the superior wear characteristics of vinyl.

Vinyl and cork tiles are perhaps the ideal do-it-yourself item because few tools and little experience is required. If you make a faulty cut, you've only ruined one tile, not the entire floor.

Rubber Flooring

Rubber flooring is used predominantly in commercial buildings. It has a high-tech look and comes in a variety of surface patterns: raised circles, raised dots, grids, bars, and leatherlike textures. It is resilient, slip resistant, and withstands cigarette burns. Some rubber flooring requires no waxing. It wears like iron and is easy to care for. It comes in the form of tiles and is therefore a candidate for do-it-yourselfers. The only real drawback is its high cost, close to the most expensive sheet vinyl.

Poured Floor

Another alternative is a poured floor, which involves laying down a substrate of plywood over which a coating of epoxy or polyurethane resin is poured. These floors are seamless and to an extent self-

leveling. Vinyl chips, acrylic flakes, or other decorative materials can be sprinkled on the wet floor, creating an attractive surface pattern and giving the surface a nonskid quality. Generally, poured floors are installed by special seamless flooring contractors, although I have installed one myself and do not think the job beyond the skills of a competent do-it-yourselfer.

Wood Flooring

Wood has excellent qualities as flooring: it is softer and quieter to walk on than ceramic tile; it shares vinyl's advantage of being relatively warm to the touch; and it's good-looking. Unfortunately, be-

Small tiles are both flooring and base of this walk-in shower and bathing enclosure. Note the windows in the tub/shower room — frosted below, they can open for ventilation above.

cause constant inundation will stain, crack, and warp most species of wood, we do not recommend it in any bath with the exception of the powder room, which gets little abuse.

Despite this warning, many people wanting the warm look and tactile reassurance of wood under their feet will use wood anyway, so the following advice is in order.

The most popular wood flooring is oak, because it is hard, quite durable, dimensionally stable, and reasonably priced. Maple is also popular, as is ash. A number of exotic woods, such as Brazilian cherry, purpleheart, and bubinga, are also durable. Woods such as pine, cherry, and walnut are too vulnerable to be candidates for the bath.

While color and pattern varies among species, it also varies within species. Not all red oak, which is often used for kitchen cabinets, looks the same, and red oak has a different tone and grain from white oak. Also, bear in mind that consistency of color varies with the wood's grade. Clear wood, which contains no knots, is highly uniform in color, whereas lower grades contain a mix of colors. While the clear grade is the most costly grade in any species, price is also influenced by other factors, such as the length and width of the stock.

If you go the wood-flooring route, as I have in several of my own bathrooms, be sure to seal the wood very thoroughly with three to five coats of high-quality polyurethane. Inspect the finish periodically to check for worn spots or breaks in the space between the boards. If yours is a household with small children, in particular little boys, you must prevent the urine from permeating the wood and causing odors.

"Floating" Wooden Floors

Prefinished laminated flooring, or *floating floors,* is one of the newer varieties of wood flooring to appear on the market. It is a flooring system consisting of plywoodlike laminated wooden planks with a veneer of the finish wood species bonded to the top surface. The planks have a tongue on one edge and end and a groove on the other, similar to strip oak, and they fit together with very close tolerances.

The rubric *floating floor* comes from the fact that the planks are glued to each other at the edges but not fastened to the subfloor. Instead, they rest or "float" on a thin carpet of polyethylene foam. This makes them quieter and warmer than conventional strip or plank floors. They are available in a variety of finish species, some rare and difficult to obtain in strip or plank form.

Floating floors come from the factory prefinished. Once the floor is laid, the job is done. Manufacturers claim that their coating, applied in controlled conditions in the factory, is tougher than coatings applied at the job site. This claim might be true, although the coating is broken where the planks are glued together, allowing an entry point for water. Also, the coating appears to be much thinner than the three coats of polyurethane we normally apply, by brush, to plank or strip floors.

A floating floor might be appropriate for a formal powder room, but for a main-line bath that will see a lot of water and traffic, it is not a good alternative.

Sleek wood flooring and paneling play well with both antique and modern fittings.

Cabinets: Custom Versus Stock

In all but the most minimalist of baths, you will need some sort of vanity and perhaps additional storage devices — which is to say, cabinets. You'll find that your first choice is betwen *stock cabinets,* mass-produced in a factory, and *custom cabinets,* which are hand-built for a particular job. The biggest difference between the two is the range of choices available to you. To realize efficiencies of scale, stock manufacturers limit their production to the most popular sizes, colors, and features.

Many stock cabinets are available in wood or are surfaced in high-pressure laminate. Stock cabinets are available in a few of the popular colors, whereas custom shops, if they build laminate cabinets (some build only in wood), offer a broad selection of colors. As a rule, stock cabinets are less expensive than custom, although some of the highest-quality stock cabinets sell for the same as custom. If you're on a tight budget, stock cabinets help you keep it. (I should note that for my laminate cabinets, I use a commercial cabinetmaker who builds for department stores, hospitals, and boutiques. His operation is so efficient that he can offer his product at the same or even slightly lower cost than average-quality stock cabinets. Perhaps there is such a shop in your area.)

In addition to being attractively priced, stock cabinets are available quickly. Retailers such as home-improvement centers, lumberyards, and kitchen dealers either have the cabinets on hand or can deliver within a week or two from a warehouse or factory.

Shopping for custom cabinets is a little different. Home-improvement centers and lumberyards are seldom a source; kitchen design stores usually are. Typically, the store's in-house designer collects your measurements, designs the cabinets, and orders them from a large factory. There are also a number of small custom cabinet shops and even skilled carpenters who will build cabinets.

A custom cabinetmaker can tailor the casework to the dimensions of your room and your particular needs and taste. If you want the lavatory height higher or lower than normal, the custom

Even though the space is restricted, this vanity's casework is designed for maximum — but separate — storage for both partners. Tile laid on the diagonal makes the counter seem a little bigger than it really is.

shop can easily oblige. Customers often want their counters higher than the customary 36-inch height, a standard established years ago when Americans were shorter. At "This Old House," we often build counters 37 inches high, which many people find more comfortable.

In a small or oddly shaped room, custom cabinetry may be essential in order to obtain enough storage or to mask the shape of the room. Indeed, one of the great advantages of custom cabinetry is that it can seize every square inch of potential storage, while making an elegant design statement in the process.

Custom cabinetmakers earn their pay, in large part, from their attention to aesthetics. Part of their job is to blend the cabinetry into the bathroom as a whole. Where the cabinet stops at a window, for instance, the cabinetmaker can provide transitional elements such as shelves for flowers or special objects, or can design custom moldings for bottom and top edges of wall cabinets.

Yet a third category of cabinets offers more flexibility in design and materials than stock manufacturers but less than a true custom shop. Called *factory custom*, these cabinets can be customized within a range of options of sizes and materials. Because they are built in a high-volume factory, where certain economies of scale apply, the cost of these cabinets can be substantially less than custom, with little or no sacrifice in quality. In the last few years, the variety and quality of factory custom cabinets has increased dramatically, and we expect this trend to continue.

Just as in baking, cooking, and winemaking, you cannot get a quality result with anything but the finest ingredients. At "This Old House," we have learned the hard way, both in our projects for television and our own projects at home, not to skimp on the quality of materials. The other side of the coin, though, is don't go overboard. The mere fact that you can afford Carrara marble does not in and of itself mean the bathroom will look like a million bucks. Good design and sensitivity to material properties can give you a high-quality result with standard materials.

●

The Family Bath

This rambling, informal family bath has room for plenty of kids, ample storage, and comforting period details, like the pedestal sinks. (Notice how the dividing wall also hides the plumbing.) A whirlpool spa is just out of sight to the left.

It is within our parents' lifetime that many American families did not even have a bathroom in their homes. The "bath" referred to that elaborate Saturday afternoon ritual during which the whole family cleansed themselves with sufficient assiduousness to last the coming week. In the 1920s, a number of American families owned an automobile but not a bathtub.

These days, the bathroom is almost as central to the life of a family as the kitchen. Most of us expect to bathe at least once a day, and we also expect to perform our prework or pre-evening-on-the-town ablutions in the privacy of our own bath. The number of bathrooms in American houses has not been pinned down with any precision by the census, but it most likely exceeds 150 million, meaning that there are about 50 million more bathrooms in the United States than there are houses and apartments.

I am the oldest of six children. For most of my childhood my family lived in a variety of houses that were too small for us. Usually, the boys shared one bedroom, the girls another, and my parents the third. Many of our houses had only one bath, sometimes two, so the "family bath" was literally that: the bath serving the whole family. While most houses these days have more than one bath, the family bath is still the paradigmatic bathroom, a room that is likely to be used by everyone in the family.

The family bath illustrated here embodies some of the basic principles of bath design, construction, and materials, which are:

- Versatility: The room must serve a variety of uses, from bathing children to accommodating visiting guests.
- Durability: All materials must be selected for long life, have good but neutral looks, and be able to stand up to hard use.
- Serviceability: The room must be easy to maintain.
- Compactness: In most houses, space is an issue, so the family bath cannot be any larger than it needs to be.
- Storage: There should be as much storage as possible, either in the bath itself or in the general vicinity, for towels, linen, medicines, cleaning supplies, and children's bathtub toys.

(Left) Family-style bathroom in a restored 1789 merchant's house in St. Augustine, Florida. The tub, which back then needed to be portable, was not exactly spacious.

Here are some design strategies for renovating the second-floor family bath of our concept house. The plans proceed from the least to the most expensive.

Scheme 1:
Minimal footprint, simple cabinetry.

Scheme 2:
This bath is only a foot and a half wider, which affords the opportunity of placing storage cabinets across the vanity wall.

Scheme 3:
Eliminating the storage closet provides room for double vanities. A skylight brings ample light and natural ventilation into the room.

Scheme 4:
This arrangement gives more space to the vanities, and could allow for a window on the outside wall.

Scheme 5:
By robbing more space from the master and second bedrooms, there is room for double vanities, more linen storage, and a window.

Scheme 6:
A skylit edition of version 5, with toilet in a separate compartment behind a pocket door.

All of this amounts to a tall order, considering that many of the requirements are mutually contradictory. The famous American naval architect John Alden is said to have asked his clients to list every tool, spare part, and kitchen utensil they planned to stow aboard before he would set about designing the interior of their new yacht. Renovating a bathroom is much the same — space is finite and every square inch must be used to its full advantage.

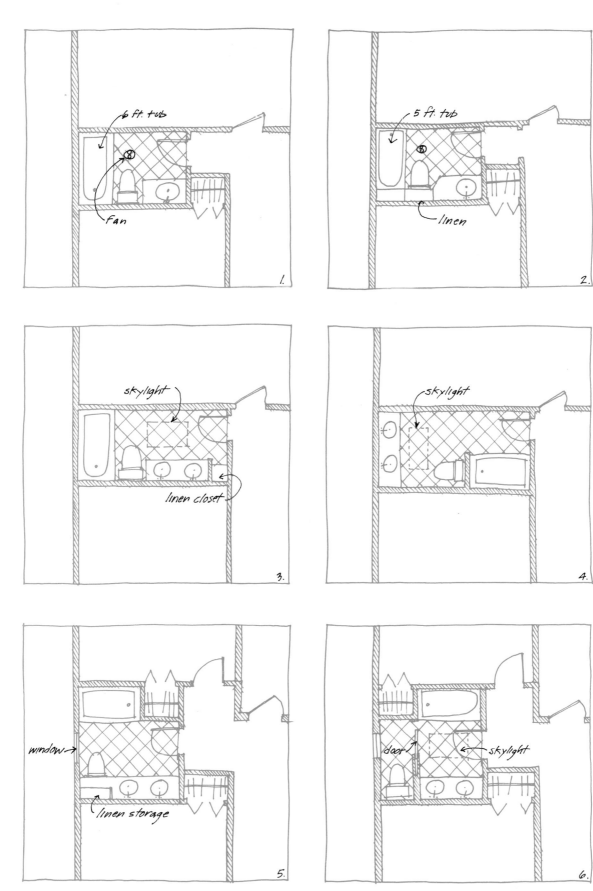

Family Bathrooms

(Left) The "family bath" aboard the 43-foot cutter-rigged motorsailer *Congar.* Like all shipboard "heads," it is a study in compaction, with toilet, sink, and shower all sharing the same closet-size space. (Proper yachting etiquette dictates that one remove the toilet tissue to a dry place before showering.)

(Right) A spare, elegant bathroom. The large bank of cabinets, generous tub, and large metal-framed window all give a sense of large scale to the small room. Cabinets do not project the full depth of the lavatory, thereby alleviating the stuffed feeling typical in a narrow bathroom.

Storage

You, like Alden's clients, might start your design process by making a list of all the things you keep in the bathroom (or would like to keep if you had the space): towels, toilet paper, cosmetics, soap, medicines, mouthwash, toothpaste, tampons, hair dryer, curlers, and so on. Expecting your storage needs to be adequately served by a free-standing chest of drawers just isn't realistic. The most efficient mode of storage is to build it in.

The prime storage area is in the vanity. It is also the lavatory and the vanity cabinet that make one of the major design statements

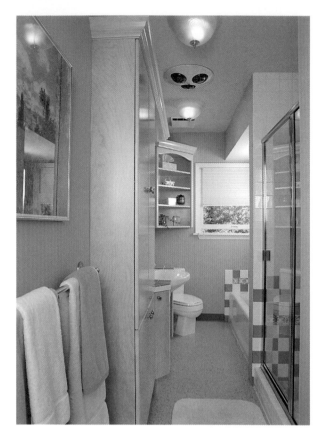

in the room. A vanity just large enough to house the lavatory is what one typically sees in bathrooms, but since this cabinet must also house the feed and waste pipes for the sink, organizing the storage below is usually a bit of a problem. Often, the best solution for storage is to build the lavatory (or lavatories) into a wall-length run of base counter. A pedestal sink is usually a poor choice for a family bathroom, since it completely eliminates any potential for storage.

Kitchen and bath designer Glenn Berger likes to make even more room by keeping the compartment beneath the lavatory as narrow as possible, while allocating maximum space to drawers. In his own house, he put only two small doors but *ten* drawers in a 6-foot-long vanity. If you have small children, he recommends equipping some drawers, but not all, with locks. "If there's one drawer and it's locked, you've got a frustrated child," he claims (from experience). "But if you have ten drawers and only six are locked, he'll be delighted to find the four that operate."

●

You will, of course, want mirrors above the sink and above any other area where applying makeup is done. Keep in mind that mir-

When I first looked at this family bath, I saw only the idiosyncrasies — the 1950s styling and the pedestal sink surrounded by counter (a maintenance headache). But then I saw the strong points: good use of a narrow space with ample storage in cupboards and a corner hutch, a separate bath and shower, good natural light and ventilation, an easily maintained linoleum floor, washable wallpaper, and a tiled bath surround.

rors, while not cheap, are a very efficient way to increase the illusion of space in the room. In the family bathroom of my own house, I extended the counter in a narrow shelf over the toilet and filled the space above the counter with mirrors. While the room is of minimal dimensions (6 x 9 feet), one never feels cramped.

For shaving, putting on makeup, or primping it is essential to be able to get close-up frontal and profile views of one's face, so locating a mirror to one side, perhaps on the door to a medicine cabinet, is a good strategy. A full-length mirror is also handy, although finding space for it inside the bath is usually impossible. Perhaps one could be located on the outside of the bathroom door or in an adjacent hall.

Since several people are likely to use a family bathroom, it's a good idea to have plenty of hooks or bars for towels and washcloths, some of them within reach of the tub or shower. These accessories come in a profusion of designs, colors, and levels of quality. When making your selection, remember that, when installed, they make a big design statement in and of themselves, so choose carefully. If you select metal, it should be good quality chrome, brass, or stainless. Some nicely designed and very colorful plastic accessories are on the market now, too. These accessories will be on display for a long time, so it's worth paying the price to get materials that will retain a handsome finish.

Tubs and Showers

Another fixture with big visual impact is the tub and shower. In our concept bath, we chose a combination tub and shower simply because that combination offers the greatest versatility at the least cost.

Tubs and tub-shower combinations are available in several materials. The traditional one is cast iron; it is durable and very smooth with a baked-on porcelain finish that will look good for many years. Cast iron does have some disadvantages. It is a very hard material and is unforgiving if one happens to fall in the shower (although the bottoms of the new tubs are etched with an excellent, nonskid finish). Cast iron is a good thermal conductor, which means that in winter it is cold and draws the heat out of the hot water in the tub. It is heavy and therefore difficult to install, and the joint between the tub and the shower surround must be carefully caulked to remain watertight.

Yet for cast iron's disadvantages, it is so durable as to be considered permanent, and in a family bathroom where sandy children and muddy dogs will be washed in the tub, it is the material of choice.

There are steel tubs on the market, but they are not recommended, as they chip, dent, may deform under the weight of water
and bather, and may rust through at the drain and faucet holes.
However, a new type of steel tub has been developed wherein a
high-quality steel shell is coated with baked-on porcelain, giving it
the appearance of a cast-iron tub. The steel skin is backed by a
lightweight epoxy, which makes them light but rigid. The thermal

(Left) A handsome and practical bathroom where design, not expensive materials, makes the room. The counter extending over the toilet helps pull the room together; a wood-paneled mirror wall lends warmth and a sense of depth; tile extending up the side walls is easy to maintain; the crisply detailed skylight and light well more than make up for the lack of a window.

(Right) Good use of color, clean lines, a nice mix of materials, and simple detailing make this a very pleasant bathroom. The only flaw is the window on the shower wall, which invites maintenance problems by allowing water a way to penetrate the wall.

properties of the epoxy are such that water stays warmer in the tubs. Installers love them because they combine high quality with ease of installation. These units are an excellent alternative to cast iron.

Acrylic and fiberglass tubs, many offered as one-piece models with integral shower surrounds, have become increasingly popular in recent years. They cost significantly less than a comparable cast-iron or epoxy-backed steel tub, especially considering the additional cost of a ceramic-tile surround. Fiberglass is softer than cast iron or steel, a quality that would be welcome in the case of a fall, and also lighter and therefore easier to install. Plastic is a better insulator than metal so is warmer to the touch. The one-piece tub-shower units have no seams to leak and most have fairly gentle, easily cleaned contours.

Plastic tubs have two disadvantages. The first is that they are not as durable as cast iron. The second point is purely aesthetic: some people want the look of cast iron and tile, which is perceived as higher quality than fiberglass or acrylic. They also like the solid feel of these materials in contrast to flexible plastic. If you plan to sell your house in the next several years, these may be considerations.

In my own house, I chose cast-iron tubs with ceramic tile surrounds. Because it is an old historic house, I didn't feel that fiberglass suited the style of the building. Also, I had my eye on future resale value. However, I have nothing against the plastic tubs and in a family bathroom that will see moderate to light use, they provide an alternative. No matter what material you choose, make sure the tub has a good nonskid texture.

Shower Doors and Curtains

For showering in the tub, you will need some way to keep the water contained, and for this purpose you have two alternatives: a shower curtain and a shower door. Each, naturally, has advantages and disadvantages. Shower doors are available in many styles, sliding or folding, and range in price from a couple of hundred to more than a thousand dollars. To provide a foolproof barrier to water splashing on the floor and walls, one merely has to close a shower door — a shower curtain must be deliberately drawn and placed to provide a good seal. Some sliding doors can be lifted out of the track for cleaning, an attractive feature.

On the other hand, the installation of a shower door does little to make the tub enjoyable for bathing, whereas a shower curtain can be drawn fully out of the way. Also, the presence of doors makes cleaning the tub and shower area more difficult. Since most doors slide on a track installed on the top rim of the tub, leaning over the track to wash small children — or the tub itself — is a nuisance.

For my own family bathroom, primarily used by children and occasionally by guests, I installed a shower curtain. At some later date, when my son can shower himself, I will install a shower door.

Tub and Shower Controls

Renovation offers you the opportunity to change from a dual-lever control system for the hot and cold water in your tub and shower to a single-handle, pressure-balanced antiscald valve. Not only does this control device offer faster, more accurate control of water volume and temperature, but a mechanism inside the valve regulates the ratio of hot to cold water at *all* times, ensuring that when someone flushes a toilet or flips on the washing machine elsewhere in the house you are not suddenly scalded with hot water. "This Old House" plumbing and heating expert Richard Trethewey strongly advocates the universal use of antiscald valves. They are required by plumbing code in some states, but, surprisingly, not in others. Pressure-balanced antiscald valves are commonly available and

(Above) This brightly colored hardware is easily manipulated by hands of all ages.

(Right) A well-detailed tub and shower surround. The stippled glass window affords ventilation with privacy and is high enough to be out of the shower's splash zone.

are not significantly more expensive than standard valves. That they are not universally required by code is bewildering.

Floors

Yet another element in a family bath that will make a big impact on the looks of a room is the floor. Although there are a number of choices for bathrooms in general, you must carefully consider the design criteria for this particular bath: versatility, durability, and ease of maintenance. Consider that the floor of your family bath will

In this recent addition to a log cabin in Aspen, Colorado, the wood stove is the most important feature of this bath. It gets the place toasty in a matter of minutes. A hall chair holds the waiting family members and their towels!

likely get inundated with water from baths, showers, and children's water-play in the sink; will get coated with dog hair, mud, and sand; and will be splattered with everything from toothpaste to urine from little boys still learning to aim correctly. You're talking about the need for a bulletproof floor in this room.

Vinyl

Vinyl is one of the most popular materials for the bathroom floor. It is inexpensive, warm to the bare feet, has good cushion, and some patterns offer adequate slip resistance. Vinyl is impervious to water except at the seams. In the bathroom, we like to install it with no seams if at all possible and to give it a self-formed cove base. In this installation, the edges of the flooring make a 90-degree turn, going up the first few inches of the walls. This reduces the risk of water getting into the joint at floor and wall and also makes the floor easier to mop. Since the amount of vinyl you will need for a bathroom is minimal, insist on buying high-quality goods.

Ceramic Tile

For durability, good looks, variety of color and pattern, and ease of cleaning, it is hard to beat tile. It comes in various sizes, the most popular for bathrooms being 1- and 2-inch-square mosaic tiles. Many tiles are available in sheets, making them easier to install. Ceramic tile can be slippery, but a couple of factors help to relieve the danger. First, the grout between tiles gives the floor some "tooth," a good contrast to the smoothness of the tile surface, and second, the grout absorbs some of whatever water falls on it, so puddles don't stay as long as they would on a vinyl floor.

In the past, the greatest objection to tiles was not the tiles themselves but the grout joints, which tended to get stained and dirty. This problem has largely been eliminated by several new types of products: epoxy grouts that are highly stain resistant; clear, silicone sealers, which effectively seal both tile and grout; and new, spray-on tile cleaners, which quickly remove any stains that might appear. If your floor is going to get heavy, constant use, you might choose a dark-colored grout to hide the dirt. The grout lines will also be less noticeable if you choose a large tile size.

A tile floor should have a tile baseboard, called a *sanitary base*, to protect the juncture between floor and wall. This will make the floor easier to clean as well.

Wall and Ceiling Surfaces

If you are gutting your bathroom, this is a golden opportunity to install a really good insulation and vapor barrier system as described in chapter 2. If your current bathroom is cold and drafty, consider installing such a system even if you don't need to remove the walls.

A very pleasant and inexpensive bath, given style with the vinyl wallpaper and patterned curtain. Sometimes, existing features — such as the yellow tile — can be used as a starting point for redesign.

You may actually save money by avoiding the need to alter the room's heating system.

In choosing the actual wall surface for your family bath, you have to balance several factors: cost, durability, ease of cleaning, and looks.

In terms of cost, the cheapest way to go is a drywall or blueboard-veneer plaster, painted with a high-quality oil or latex enamel, either gloss or semigloss. Next on the cost scale would be the same surface coated with a high-quality, water-resistant, scrubbable wallpaper. I did this in my family bath, and it has held up well for five years at this writing. Most expensive, and also most durable, cleanable, and maintenance free, would be tiled walls.

I think that if I had my family bath to do over again, I would cover walls and floor with tile, installing a sanitary base molding at

Neither a large nor a luxuriously appointed bathroom, yet the mirrored wall and the clerestory windows give it light and a pleasing sense of space.

the junction of wall and floor. Tiling is something you can do yourself — with patience, a little practice, and the right tools.

Illumination: Windows, Lighting, and Electricity

Natural illumination is usually desirable in a bathroom, especially a family bath; being exposed to the neighbors is not. These requirements, combined with the tight space one is usually faced with in a bathroom, can create a quandary as to where to place a window. Ideally, the window should be in a position that allows you to use the bathroom without drawing the blinds, but with mirrors, shower surrounds, medicine cabinets, towel racks, and storage, wall space is usually at a premium.

There are a number of strategies for dealing with this problem. One would be a clerestory window located high on the wall

(Left) Since the footprint of this bath was so small, the renovation focused on making it seem bigger with mirrors, glass block between tub and sink, and a unifying horizontal band of solid-surfacing molding and checkerboard tiles.

(Right) A clever detail for catching the soggy towels.

above the shower, toilet, or lavatory. Another would be to eliminate windows in favor of a skylight. In my family bath, the window is right in the middle of the shower surround, a location that, regrettably, I did not alter during the renovation. (If I had it to do over again, I would eliminate the window in favor of an operating skylight for both light and ventilation.) Still another approach would be to use glass blocks as part of the shower surround. Finally, you have the alternative of eliminating windows altogether and installing instead good artificial lighting and an exhaust fan.

In some historic houses, like our concept house, an existing window cannot be moved or altered without detracting from the house's appearance. In this case, you have no alternative but to work around it. A clever design may even turn this apparent liability into an asset.

●

Artificial lighting should be installed over the lavatory. Bathroom designers also recommend lighting along the sides of the mirror; for without it, a man has a hard time getting an unshadowed view of his neck while shaving, and a woman has a difficult time applying makeup. Lighting may be desirable in the toilet area for reading and in the tub and shower areas. Any light fixtures in the shower must be protected against moisture. (For more on lighting, see chapter 10.)

In a family bath, you may consider accepting the less desirable quality of fluorescent lighting — poor color rendition — for the desirable qualities of ample, shadowless light and efficient operation. These qualities may be especially welcome in a household with children — excellent light to bathe them by, efficient operation when they forget to turn the lights off.

In a family bath, it is probably not a bad idea to install two lighting systems: a source of ample, general illumination used when bathing the kids, cleaning the bathroom, or stocking the linen closets, and task lighting at the mirror, perhaps a row of makeup lights above or to the sides.

●

Your bathroom renovation may also call for increasing the number of electrical receptacles in the room. We use many more bathroom electrical appliances than in the past — hair dryers and curlers, shavers, electric toothbrushes, contact lens sterilizers, and so on. The National Electric Code (and common sense) requires that one electrical outlet be placed near the sink. It must be of the ground fault interrupter (GFI) type, which will automatically shut off power in the case of a short circuit, as occurs when you try to wash and dry your hair simultaneously. Even if not required by local codes, it is good practice to run all circuits in the bathroom through the GFI to reduce the possibility of electrical shock to an absolute minimum.

●

These are some of the fundamentals of renovating what we consider to be the "typical" family bathroom. But since, in renovation, such a thing as "typical" rarely exists, and since your renovation is likely to incorporate elements from many other types of bathrooms, I invite you to read (or at least browse) onward. In the chapters that follow, you'll find much additional information that will contribute to your project.

●

Lots of light was brought into this small bath by converting the window to french doors. The railing outside is an obvious safety feature.

The Children's Bath

A whimsical collection of odd tiles and mismatched handles on the sink are any kid's prerogative.

With six kids in our family, my mother often used to wish out loud for a fully tiled bathroom with a drain in the middle, a room that could simply be hosed down at the end of the day. Now, a father myself, I see what she meant. If you are designing a bathroom to be used primarily by children, there is no reason not to create a room that kids can have fun in — splashing in the tub or playing in the sink without fear of ruining the walls or floor. There is equally no reason the bathroom cannot be designed so that children can use the toilet and wash their face and hands at the lavatory all by themselves. Even in a household with only one or two baths, incorporating "kid friendly" features into the family bath will make life easier for you, and will help to give your children independence and a great sense of pride by including them in the daily life of the household.

Most of the physical and mechanical considerations that apply to a family bath also apply to a children's bath: placement and type of lighting, heating, ventilation, windows, types of tub and shower combinations, and types of flooring, so there is no need to repeat ourselves on those subjects. But two points should get special consideration: safety and ease of maintenance.

Safety

More accidents occur in the bathroom than in any other part of the house. The dangers are especially great for children because of their inexperience, limited coordination, and their boundless curiosity. There are many hazards in the bathroom: scalding water, slippery surfaces, sharp or pointed objects, dangerous medicines, and heavy-duty cleaning substances. The dangers can be greatly reduced by sensibly designing, furnishing, and equipping the room.

Water Controls and Bath Fixtures

Separate faucets for hot and cold water are not a good idea in a children's bathroom. Youngsters are not as adept as adults at regulating water temperatures, so single-lever controls are preferable. They offer more safety in the bathtub, the shower, and the sink, too. Water controls for the bath or shower should be equipped with a pressure-balanced antiscald valve, even if it is not required by local

plumbing codes. As we discussed in the last chapter, this valve prevents the water temperature from increasing suddenly when cold water is drawn from the plumbing supply elsewhere in the house.

Another device worth investigating is a temperature-limiting faucet, which allows you to set a maximum temperature for water entering the tub or shower, further protecting children from scalding water.

In a master bathroom, it's nice to have a separate shower and bathtub, but in a child's bath this is usually an extravagance. What's really needed for bathing youngsters is a tub, and, as in the family bath, we would combine it with a shower for versatility and economy of dollars and space. We outline the choices in materials in the previous chapter, so suffice it to say here that whatever tub you choose, make sure it has a good nonskid bottom.

Whether to install a shower curtain or a shower door is a judgment call on your part. If your children are older and can shower themselves, a door, which provides a good, watertight seal upon closing, would make sense. However, if your children are younger and you are bathing them, a shower curtain provides unrestricted access to the tub. It's worth pointing out that small children

(Above) Niches for toys and a full-length mirror to admire growing bodies are good ideas here. The recessed shelf around the sink makes it easier for kids to reach the faucet.

(Right) Another likely candidate for teenagers — shelves on which to stash stuff and a shower stall that's easy to clean.

can't smash their fingers in a shower curtain, while they might in a sliding or swinging door.

Floors

By the same token, the bathroom floor should be chosen at least partly on the basis of safety. Since some water will inevitably land on the floor, it's best not to use flooring material that will be slippery when wet. Smooth vinyl flooring is a dubious choice; water tends to pool on vinyl, creating perfect conditions for a fall. If vinyl flooring is selected, it should have a nonskid texture.

Our choice would be ceramic tile or perhaps some other waterproof masonry finish such as dyed concrete or terrazzo, which in some areas of the country may be cost competitive with ceramic tile. Here again, safety and ease of maintenance are the key words. Any tile with a shiny, glazed surface would not be a good candidate for this room. Choose a tile with a good nonskid surface. An unglazed porcelain tile, typically used for the floors of shower stalls, would be a good choice. If the floor is composed of small tiles, the grout lines themselves will provide some traction.

Storage

Another safety consideration is storage. Usually, it's a great convenience to have ample storage in a bathroom, but in a children's bath, this may not be such a good idea. The danger we see is that if the storage capacity is generous, adults will be tempted to fill it with things that are best kept out of the hands of youngsters. A medicine cabinet is a dubious feature in a child's bathroom; we would prefer not to put one in unless it has a childproof lock. Similarly, a vanity cabinet could pose some dangers. There is always the possibility that cleaning solutions or other noxious substances may get stored there, and even with childproof locks, a child might get at them. A pedestal sink, on the other hand, offers no hiding places for things that cause problems.

However, the typical household needs all the storage it can find in the bathrooms, and this suggests a storage scheme that provides child-secure compartments for cleaners, medications, and toiletries (perhaps up high), along with lower storage that is accessible to kids.

Open storage, in the form of shelves or cabinets without doors, offers stowage for towels and washcloths, as well as for rubber ducks, tugboats, and other bathtub toys. There should also be an abundance of hooks where kids can hang up their own towels and washcloths. Children get a lot of things wet, so it's not a bad idea to plan for plenty of towel bars or other devices for drying things out. At the same time, it's important to avoid installing sharp or pointed hooks, towel bars, toilet paper dispensers, or other hard-

This simple bath, suitable for family or kids, has shelving, an easy-to-clean floor, and good access to the tub. Plenty of places for toys and clothes, too.

ware. Rounded corners and soft objects are desirable for any of the hardware and furnishings in the room.

Electricity

Building codes require that electrical receptacles in a bathroom be of the ground fault interrupter, or GFI type, which helps to prevent electrocution. Although GFIs offer great protection, we would still exercise restraint in number and placement of receptacles in a child's bath. Common sense (as well as the National Electric Code) dictates that outlets should be kept away from wet areas and installed at a height where young children cannot reach them. In a child's bathroom, there is probably no need for more than the required single outlet placed near the sink. As an added measure of safety, this receptacle can be protected with childproof covers.

Bath as carport. The walls and ceiling are paneled with Oronite, used for the roofs of patios and Miami carports — cheap, easy to install, and weird. Note the outdoor lighting fixture, too.

A built-in night light is a great feature, both for adults and for kids. All lights in the room should be protected with the GFI.

The bathroom should have a mechanical ventilation system in the form of a bulit-in fan to remove moisture. It is not a bad idea to place it on a timer so that it will automatically turn off if a youngster forgets about it. If there's a heat lamp, it, too, could be placed on a timer.

Locks

We would exercise caution about equipping the bathroom door with a lock. One of the most exasperating things a child can do is lock himself in the bathroom while his parents alternately try to cajole or coerce him into opening the door. The remedy is to do without a lock or to install one that can be easily defeated from the outside.

For Easier Maintenance

Floors

As we mentioned earlier, ceramic tile makes an especially good bathroom floor. Besides offering some traction, it's a surface that's easy to mop down for cleaning. The old problem of the grout getting stained and dirty can be lessened by using an epoxy grout, a latex-

This bath was added to an attic to give teenagers their own domain. Good design features prevail: skylight with blind, mirrored wall, and bold tile pattern to enlarge and unify the space.

enriched grout, by sealing the tile with silicone, or by choosing a gray or other darker-tone grout color. Sanitary base tile should be installed at the base of the walls. This will protect the bottom of the walls from any water on the floor and will make it easier to mop without dirtying or harming the walls.

Vinyl flooring is very easy to clean, and it can be formed into a sanitary base. If a wooden baseboard is used, make sure the joint between it and the flooring is well sealed with a fat bead of silicone caulk. This will make mopping the floor easier, too.

Walls and Tub Surround

Both around the tub and on the walls, it's useful — if, indeed, not essential — to install materials that are durable and easily cleaned. Ceramic tile performs well in this regard, which is why it is the standard for the heavily used shower rooms in swimming pools and athletic clubs. You have greater freedom in selecting wall tiles than floor tiles, because they will not see the same heavy wear, nor do they need to have a nonskid texture. Since the color, type, and application pattern of the tile will make the major design statement in the room, you must carefully consider the aesthetic consequences of your choice.

A stately children's bath that could be redecorated as a guest bath when the kids grow up. There is ample space and light, tub and shower are separate, and the tile makes it easy to keep clean. The built-in seat between tub and shower is a great place to dry and dress squirming kids.

An option for both tub surround and wall surface is solid surfacing (trade names: Corian, Gibralter, Avonite, Fountainhead, Surell), which provide a water-resistant and easily cleaned surface with few joints. Du Pont recommends that its product, Corian, be glued to moisture-resistant drywall, cementitious backer board, or marine-grade plywood with its own brand of panel adhesive. The joints between sheets should then be covered with narrow strips or battens of Corian. On vertical surfaces, ¼-inch-thick solid surfacing can be used (instead of the ½- or ¾-inch-thick material required for horizontal surfaces), which, considering the speed of installation, makes this wall covering competitive in price with tile.

In the tub itself, the main maintenance problem is the standard pop-up drain, which quickly gets clogged with hair. Children are apt to lose small objects down the drain, which can result in late-evening calls to your plumber. A lot of headaches can be avoided by specifying a drain equipped with an integral strainer. Though such an installation is not standard practice in residences, several models are available for commercial applications.

Bathroom Fixtures

A one-piece toilet is probably the best choice for a children's bathroom. Its simplified design results in surfaces that are much easier to keep clean than those of the usual two-piece model. The cost of a one-piece toilet is higher, but the benefits in appearance and sanitation are worth it. Better still might be a wall-hung toilet, which will make the surrounding floor area very easy to clean and eliminate one more crevice — the joint between floor and toilet — for dirt and water to hide in.

Counters

The best counter surfaces, from a maintenance point of view, are solid surfacing and high-pressure laminate, in that order. Solid surfacing is easily cleaned and impervious to water. Some manufacturers make lavatories that can be integrated into the counter, eliminating the rim seal, that vulnerable joint between sink and counter. (Du Pont manufactures five lavatory bowls that can be bonded to a counter of matching or contrasting Corian, along with two lav models molded right into the countertops.) The chief disadvantage of solid surfacing is its high cost, roughly $80 to $100 a linear foot for fabricated countertops, depending on the level of detailing, a figure several times that of a high-pressure laminate.

High-pressure laminate is far more cost competitive than solid surfacing. Laminate comes in many colors, it is easy to clean, and, if it is treated with care, can be very durable. All joints must be well sealed to prevent standing water from penetrating to the particleboard substrate below. Otherwise, the particleboard will absorb water and expand like a sponge, exploding the countertop. I speci-

Good lighting, bright colors, a shower curtain instead of a door, and easy-to-grasp faucet handles are the features of this children's bath. I'd be a little concerned about the hard corners on the tile counter, though.

fied high-pressure laminate for the countertops in three bathrooms, one of which is a children's bath, and for the laundry room as well. They have been in constant use for five years and still look new.

Tile, as we have discussed, is an excellent material in wet environments, but it is very hard, and thus presents a potential hazard if a child were to knock his head against the counter's edge. Also, the grout lines present many crevices to catch toothpaste, soap, and whatever else a child may slather onto the counter. Since the cost of tile countertops is roughly equivalent to solid surfacing,

The bath looks old-fashioned but has thoughtful and inexpensive kid-friendly features — the hooks mounted down low, the mirror on the sink, and the footstool. Towels warming over a radiator and a braided rug cozy up bathtime.

choosing between the two is what they call a "no brainer" in our opinion — we'd choose solid surfacing with an integral lav bowl.

Accessibility for Children

Because of their size, small children may have trouble using a standard-height lavatory or reaching things that would be within normal range for an adult. There are a number of strategies for making the bathroom more accessible and, therefore, comfortable for a child.

A wild and crazy bathroom with a great sense of fun. Notice the kid's and adult's sinks and, reflected in the mirror, the painted pipes for the water heater.

Fixtures should be installed at normal heights, since any extra-low fixtures will certainly become a hindrance when the child grows up or when you sell the house. To give the child access to the sink, you can provide a portable step stool, but you may consider incorporating into your design a built-in step in front of the lavatory that can be removed at a later date.

A number of other adjustable devices can accommodate your child as he or she grows:

- A pole-mounted shower head that slides up for adults and down for children. It can be detached for use as a hand-held sprayer (which is as handy for cleaning the tub as it is for bathing). Some models look like brightly colored sea serpents' heads.

- Mirror squares can be taped to the wall for use by small children and then gradually raised until the child is tall enough to see himself or herself in a standard mirror above the lavatory.

- Because a cabinet recessed into the wall behind the lavatory may be too far for a child to reach, consider designing a small cabinet or drawer at a low elevation, where your child can place toothbrush, comb, and other grooming aids.

- Towel bars can be installed lower than normal and then raised as the child gets taller.

Layout and Location

A children's bathroom doesn't require a lot of space. It might occupy the minimum for a three-fixture bathroom, 5 × 7 feet, although any additional space available would certainly help to make the bathroom more comfortable. Extra space may be needed if the toilet is to be placed in a separate compartment, as is sometimes the case when the bathroom is shared.

Most commonly, a children's bathroom opens onto a hall, although it may have two doors to serve two children's bedrooms, or to serve as a guest bath as well. These considerations depend largely on the configuration of the house, general design considerations, and, of course, the budget. Whatever its size and placement, a bathroom designed specifically for children can go a long way toward relieving the house's other bathrooms of kids' commotion and messiness.

•

The Powder Room

In adding a powder room to their Park Avenue apartment, the owners wanted to recreate the splendor of an Edwardian sitting room for the enjoyment of their frequent dinner guests. The design is taut, the materials first class, and the workmanship superb. Notice the inlaid diamond on the panel of the door, and the art deco vanity base.

Typically the powder room is a small half-bath located in the public area of the house. It is intended to be used primarily by guests and should present to them a formal oasis of privacy that they can use in tranquillity. From a renovation standpoint, it is one room that offers tremendous functionality for relatively little space and few construction dollars. This makes adding a half-bath a very attractive way to increase the value of a house both in terms of use and resale value.

Finding the Right Location

Generally, the powder room is somewhat separated from the living areas so that people can use it without attracting attention. Its doors should not open directly into the living room, family room, dining room, or kitchen. Often the powder room is wedged under the main stair or placed in an entry hall, a bedroom hall, or a utility area. Though the powder room should be a discreet distance away from where guests will gather, it shouldn't be so far away that using it is inconvenient.

Although in concept the powder room is placed near the formal rooms of the house, there are certainly no hard-and-fast rules about this. In our Weatherbee Farm project, in Westwood, Massachusetts, we put a powder room just off the kitchen. Our homeowners spent much of their time and did most of their entertaining in the kitchen, so it was the logical location. Similarly, in my own house, the half-bath is located in a foyer off the kitchen, where it is convenient for children to use but still private enough for guests.

If you're fitting a powder room into an existing house, you may not have much choice about where to put it. In one of our projects in Boston, we found space for the powder room underneath the stairway in the center hall. Out went a closet and in went a lavatory and toilet. We positioned the toilet against the low end of the staircase, with just enough headroom for a man of average height to stand there. The room is not palatial, but it takes very little space from the rest of the living area, doesn't infringe on the house's architectural character, and is just around the corner from the kitchen, the dining room, and the living room.

For very cramped locations, it's possible to order a tiny lavatory that fits into a snug corner. (In extraordinarily tight quarters, you might follow the European practice of putting the toilet in a compartment and installing the lavatory outside.) Toilets, too, come in small sizes, though we would specify a full-sized toilet if there is room for it; you do give up comfort with the smaller fixture. Building codes in some municipalities require only 21 inches of clearance between the front of the toilet and a facing wall. This is truly minimal; 30 inches would be more comfortable. Designers consider $4\frac{1}{2} \times 5$ feet to be the minimum dimensions for a half-bath.

Equipment and Decoration

Unlike a family bath, which one might characterize as a utilitarian room with some degree of polish, a powder room is a polished room with some degree of utility. We've visited powder rooms as elegant as a fine living room and as funky as the rest room in a roadside diner. There is no reason the room cannot be an extension of the character of the living quarters. Chef Julia Child has plastered the

(Above) Simply and tastefully done, two interpretations of the classic powder room with pedestal sink and wainscoting.

(Right) In a small, square space there are only so many ways to configure the sink and the toilet, but by playing with different combinations of arrangement and fixture styles, you can generate many possibilities.

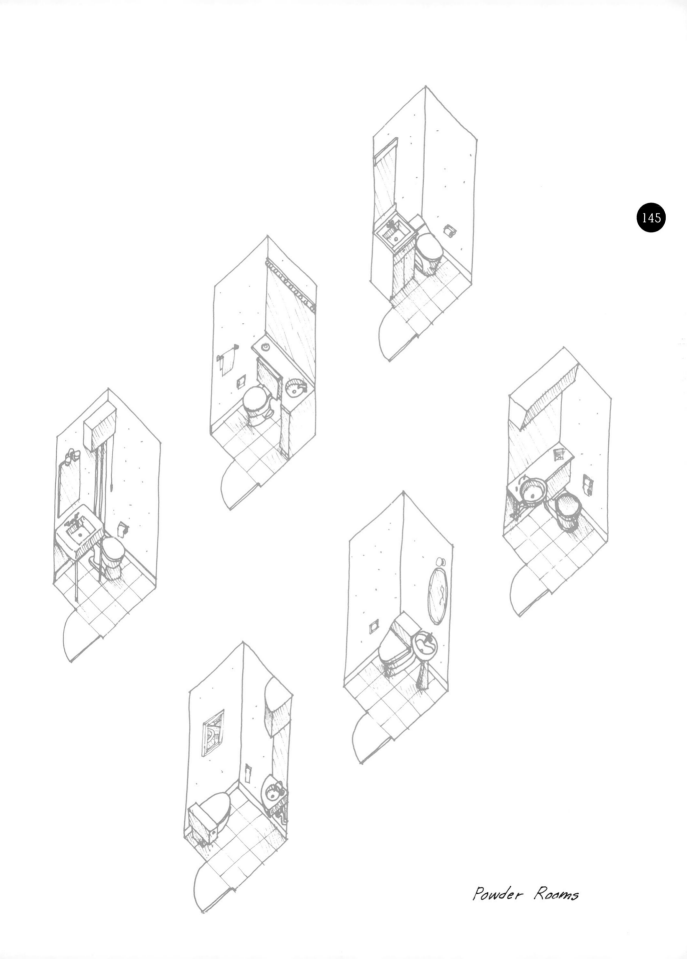

Powder Rooms

(Left) Powder rooms are a good opportunity to experiment with unusual design and materials in a confined space. Marble and glass block create a compelling sculptural element in this modern house.

(Right) A successful experiment in riotous wallpaper and a florid tile color.

walls of her powder room with French posters, so the room is reminiscent of a Parisian kiosk.

The half-bath doesn't contain nearly as much moisture as a full bathroom, so you don't need to tile the walls or worry about the effects of humidity on artwork that you might like to display there. It is the one bathroom in the house you *could* carpet, although none of us at "This Old House" recommends doing so. But just about any other finish flooring is appropriate: wood, tile, vinyl, terra cotta, and stone. In the powder room, you are fairly free to follow your feelings about style and to experiment with decorating ideas. If located adjacent to the entry, designers consider it an extension of that space, so you may want to decorate it accordingly.

Lighting in a powder room should be flattering and a bit romantic, as opposed to merely functional. A pair of sconces bracketing an antique mirror above a pedestal sink is one possibility; or, in

(Above) In renovating an apartment that the architect scorned as "a featureless entity," he enlivened this powder room by trompe l'oeil. All architectural features seen here are illusions created by painting, marbling, and wood graining.

a modern house, recessed cans or track lights that spotlight artwork hanging on the walls is another. In general, the fixtures selected should generate warm light, which enhances skin tones and makes your guests look their best. Fluorescent lighting would therefore be inappropriate. Since the quality of the light in the room depends to a large degree on the color of the walls, pink or rose-hued paint or wallpaper will generate the warm, red hues most flattering to one's appearance.

A powder room has no need for a window, but it does need mechanical ventilation. The fan is essential both for removing odors from the room and for generating "white noise" that will give your guests acoustic privacy.

Ordinarily, very little has to be stored in a powder room, so a vanity is unnecessary. An elegant pedestal sink looks right at home, as do many novelty sinks that would be inappropriate in a more utilitarian bathroom. Since the powder room is such a small

(Left) The illusion of greater space is created here by the full-wall mirror and by suspending the vanity and lighting beneath it.

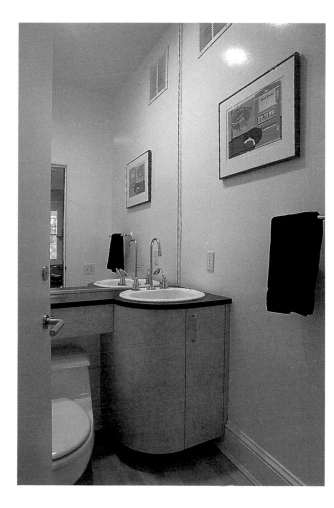

These two powder rooms enclose the required sink and toilet with efficiency and style.

and intimate space, the fixtures you choose will become the focal points of the room.

Finally, it's worth noting that not every bathroom falls neatly into a category. In the barn-style house we renovated in Concord, Massachusetts, the owners decided they wanted the powder room equipped with a shower, because the library on the first floor would occasionally double as a guest room. No hard-and-fast rules govern the design of powder rooms, only some general guidelines to help create a facility that will serve its purposes well. There's plenty of latitude for an independent approach — your goal is to satisfy your own household's needs and tastes.

●

(Top) Attic space converted to guest sleeping loft/bath.

(Bottom) This Sonoma Valley guest bath is welcoming and unfussy.

(Above) The gentleman's ranch guest bath — and not for the ranch hands. Twigs on the vanity doors are salt cedar, a material we used in our Santa Fe adobe project.

(Top right) In this small guest bath, the owner's collection of old prints and drawings plays to the "men's club" feel of the old-fashioned marble counter and faucets.

(Bottom right) In this third-story guest suite, old fixtures were installed to give it the feeling of "already there." The resulting bath is simple and spare.

Louvered doors hide the laundry center at the far end of this skillfully designed family bath.

The Utility Bath

If in any bathroom form should follow function, it is in the utility bathroom. The name says it all. This room is pure workshop, a place to efficiently wash, dry, sort, and iron and perhaps to store clothes and linens or to wash oneself or one's children or pets with not a shred of concern about being dainty. The room need not — indeed, ought not — be fancy; it should be logically designed, easy to maintain, and all components should be sturdy and straightforward, of clean line and "industrial strength."

One concept of this room is as a kind of mudroom utility area, in which one sheds dirty gardening overalls, swim trunks, foul-weather gear, mechanic's coveralls, and other soiled clothing where they will be washed, then showers (or bathes the kids) in an adjoining bathroom before properly entering the house.

Another concept is a laundry room on the second floor or elsewhere in the house, which contains a vanity, toilet, and perhaps a shower, not so much because this is the ideal combination of rooms, but because it offers the opportunity to expand the house's bathing and toilet facilities while at the same time gaining laundry space.

Location

Where to install a utility bathroom depends on several factors: your house's layout, the space available, the size and age of your family, and when and how you intend to use the room. There are two general strategies:

A common one is to put the laundry facilities close to the bedrooms, since that is where most of the laundry originates. Dirty garments, bedclothes, and linens can easily be brought from the bedrooms to the laundry and back again. If you tend to do your wash at a time of day when you are habitually in the bedroom area, such a location might be ideal. But if a housekeeper, nanny, or cleaning person does your laundry, you may not want outsiders near your private bedroom sanctum. And if a laundry room is likely to be used in the evenings or on weekends, ask yourself if you want the noise of the washer and dryer, and the activity of what amounts to a workshop, close to the space in which you sleep and relax.

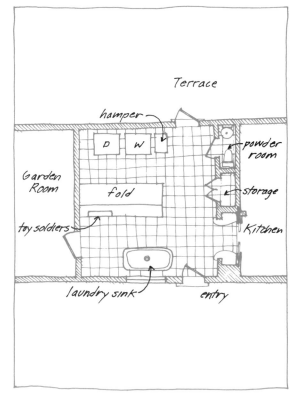

Terrace

hamper

D   W

powder
room

Garden
Room

fold

storage

Kitchen

toy soldiers

laundry sink

entry

A lot of thought went into this mudroom/laundry/ potting room/bath. The owners confess it is the most used room in the house!

If your bedroom wing is on the second floor, and you are typically downstairs when you do the wash, then locating the laundry room near the bedrooms means you will have to run up and down stairs each time you want to change loads or perform some other wash-day function.

Another approach is to place the laundry area close to the kitchen. True, you will then be committed to lugging laundry up and down stairs, but such a location does offer some offsetting advantages. First, it's easy to switch between doing the laundry and working in the kitchen. Second, you may be able to position the room to serve as the mudroom/utility/bath described above. Finally, in general space-planning terms, the utility bath is now grouped in a public work area of the house, which gives the room a different psychological spin than if it is among the bedrooms.

When I was a kid, most houses were built with laundry chutes. Several of my family's succession of residences were equipped with one, and I remember the delight with which my brothers and I would place ourselves, each other, or other inappropriate or hazardous articles in them for the ride to the first floor or basement laundry room. A laundry chute is still not such a bad idea, for it gets wet or soiled clothing out of the bedroom area and eliminates at least one leg of the down-and-up trek with the laundry basket. In an extensive renovation, it may be possible to reinstall one, perhaps integrating it with a plumbing or electrical chase.

●

When I renovated my own house, an 1836 Colonial Revival, I wanted to put the utility room on the second floor, but we simply could not find the necessary space without cutting into the period bedrooms. Thus, the utility room went downstairs into space formerly occupied by central chimneys that had been taken down at the turn of the century. My wife and I find the downstairs location very handy. It keeps the laundry activity away from the master suite and near the working portions of the house, and it is easily accessed while in the kitchen, backyard, or relaxing in the living room or TV room. The drawbacks? At only 8 × 10 feet, it is not really big enough to accommodate the sorting, folding, and ironing of clothes; and it is adjacent to the living room, so the noise of the washer or dryer intrudes on the tranquillity. Finally, the room has a sink but no shower, which would be really useful. (There is a half-bath across the hall.) In the next house I renovate, I will endeavor to create a laundry/bath/mudroom suite, off the utility entrance, in the working portion of the house.

In the house shown on page 156, the ideal location for the utility bath would have been closer to the back door; but with the

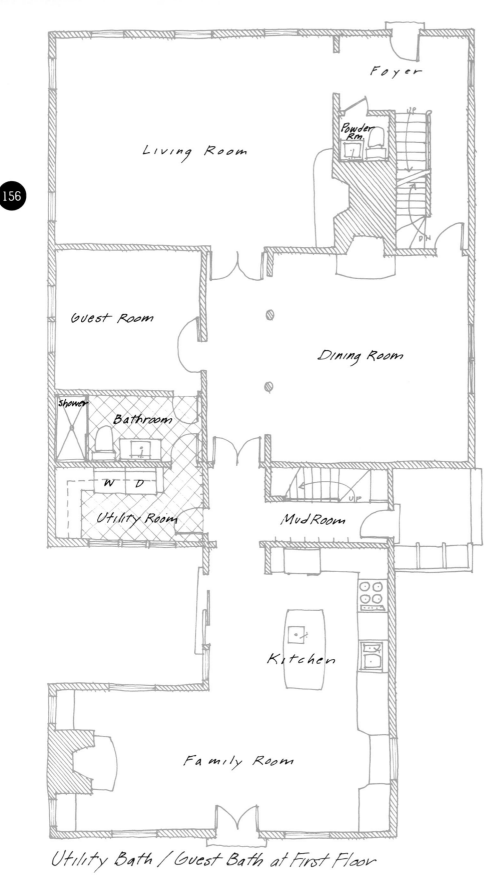

In our concept house we explored ways to install a utility bath. On the first floor we created additional space by pushing the back wall of the guest room into the courtyard, but clearly one could sacrifice the bedroom, thereby avoiding the expense of the addition.

The door between laundry and bath gives greater access from the family room yet allows the bath to be closed off for private use by a guest.

Utility Bath / Guest Bath at First Floor

(Right) A good separation of bath from laundry, with guest access and kitchen access.

(Far right) This bath functions both as powder room and utility bath, although it lacks bathing facilities. The guest room becomes a study.

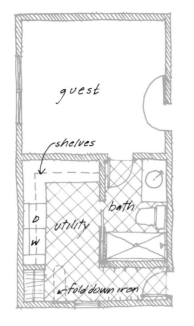

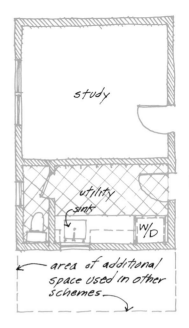

(Right) This plan maximizes the space available, and has enough room for a potting area. The study can still work as guest room.

(Far right) This plan frees up even more wall space by retaining the shower and eliminating sliders.

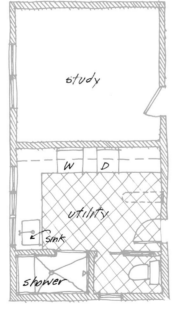

Utility Baths

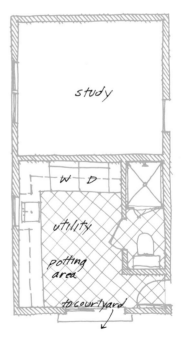

stairs and kitchen flanking the entrance, that was impossible. The next best location was across the central axis, where we created a utility suite, the bath portion of which could serve the library/guest bedroom and could even double as a powder room. For a typical family house — i.e., one with active kids generating lots of laundry, a great deal of family time spent in the kitchen preparing and eating meals, and most entertaining done casually — we think the placement and layout of the utility bath is an excellent compromise.

A compact first-floor laundry adjacent to a small bath — great for a kids' mudroom, and easy access to the john when you're busy playing outside!

Organization of the Space

Once you have sorted through the general space and planning issues to determine the location of the utility bath, it is time to start mapping out its internal geometry. The first thing one is tempted to do is to save money and space by combining the two rooms into one. This is, in fact, what "This Old House" did in the very first project, a dilapidated mansard-roofed Victorian in the Dorchester neighborhood of Boston. The adventure produced mixed results.

We had two second-floor bathrooms to install on that project, one a master bathroom hidden away in an extension of the master bedroom; the other a family-type bath to serve the other bedrooms. We decided to locate this room right off the stair hall, convenient to anyone in the house. And since we were on a tight budget, we thought that combining the laundry facilities with the bathroom would be a reasonable way to cut renovation costs and use the limited space as efficiently as possible by concentrating all the plumbing in one room. Besides, we reasoned, if the space was convenient to the bedrooms as a bathroom, it would be convenient as a laundry room as well.

Accordingly, we installed a full bathroom containing a bathtub, toilet, and double-bowl lavatory. The room was relatively long and narrow, so at one end we installed a side-by-side washer and dryer. Such an arrangement was the only practical way of installing both a second full bathroom and the house's laundry facilities on the second floor.

The project did save some money by combining functions, but anyone doing the laundry would intrude on the privacy of anyone else hoping to use the bathroom. The lesson we learned from this and subsequent projects was that to combine the two rooms compromises both, and we have urged our homeowners to separate them. Recognizing that most home renovation projects are absolutely governed by both financial and space constraints, such a separation may not always be possible. Still, it is worth trying very hard to locate at least the toilet, if not the whole bathroom, in a separate compartment.

Equipment

To be a functional workspace, the utility room must be well
equipped. The basic laundering equipment is the washer and
dryer, each of which measures roughly 27 inches wide by about 30
inches deep, depending on the make and model. Side-by-side machines have always been the most popular, but floor space can be
saved by using models stacked one on top of the other. Stacked
washer-dryers still suffer somewhat from a reputation for inferiority
because some models were down-sized, and seemingly down-engi-
neered, for apartment use. These measure roughly 24 inches wide,
24 inches deep, and 6 feet high, but White-Westinghouse manufactures full-size, robust-quality machines that can be stacked or installed side by side. Both machines are, or course, front loading,
which means you have to bend to use the washer, which is placed
on the bottom, but the dryer is at eye level. Side-by-side machines

**A sliding door closes off
the master suite from the
laundry area.**

are easier to use, but often in a small house or condominium one does not have the luxury of space to accommodate this arrangement, so "stackers" are an attractive alternative, and one to which I have resorted in several renovations.

Another basic component of the laundry bath is a sink or lavatory. Here again, form should follow function. Although a proper china lavatory may be appropriate to a bathroom, what you really want in a laundry is a wide, deep sink. A stand-alone laundry sink is the traditional (and ugly) standard, but you can use a deep, single-bowl kitchen sink mounted in a vanity and strike a balance between function and aesthetics. A kitchen sink has the additional advantage of accommodating a sprayer, which is extremely handy in a laundry room. You can improve the sink's utility by equipping it with a high curving spout for ease in filling buckets, watering plants, and hand washing articles of clothing.

Storage is absolutely essential — the more the better. The cupboards, cabinetry, or other storage devices should be large enough to accommodate the largest boxes of detergent and bottles of bleach you use, not to mention the paper towels, toilet paper, shoe polish, sea boots, and other items that will inevitably find a home in this room. If you have small children in the house, it would be prudent to devise a storage area that can be locked, to keep harmful substances out of their hands. Hampers or other storage would be useful for collecting and organizing the items to be washed. The hampers could be subdivided to separate whites and colored items. If you plan to store clean linens and towels in the laundry, consider wire or plastic baskets that slide into the cabinets, or perhaps built-in high-sided, roll-out shelves.

For cabinetry, it is hard to beat the cost, durability, and ease of maintenance of particleboard covered with high-pressure laminate. There is seldom a need to have these cabinets custom built, because high-quality stock units are widely available. However, I purchase my laminate casework from a shop specializing in commercial applications. Since they do nothing but laminate, this shop is able to produce high-quality cabinetry as cheap or even cheaper than stock. Perhaps you can find such a shop in your area.

There is nothing wrong with using open shelves for storage. Plastic or epoxy-covered wire grid shelving is now widely available at just about any home center. It is good-looking, fairly inexpensive, and can be installed by a person of average handyman skills. The openness of the grid is great for allowing linens or towels to breathe and also lets you see what is on the high shelves. Another possibility is standard plywood or pineboard shelving. It should be primed and then painted with a high-quality gloss enamel to make it easily

cleanable. For added flexibility, use adjustable hardware. The standards should be stout and the stanchions well anchored to the wall studs, not merely to the plaster or gypsum board.

A place to fold the laundry after it's been washed and dried is essential. If none is provided, the laundry will end up getting sorted and folded in another room — in our house, that's usually the formal living room. Space at all permitting, you will find it a great convenience if you equip your laundry with some sort of countertop work surface. This might be a stationary top, a pull-out surface, or a fold-down device. The work surface should be at least 2 feet wide, and if you can place it next to the dryer, you will gain some additional area. As in the kitchen, more counter space is better.

A rod, pole, or rack should be provided for holding clothing on hangers. Some articles may need to hang to drip dry, and clothing destined for storage in closets will be easier to organize if the individual garments can be hung together in the laundry area before being transported to their final destinations.

Some facility for ironing will be needed, or else that activity, too, will end up being conducted in the living room or in front of the

This bath contains a laundry module that can be closed off with a sliding door.

163

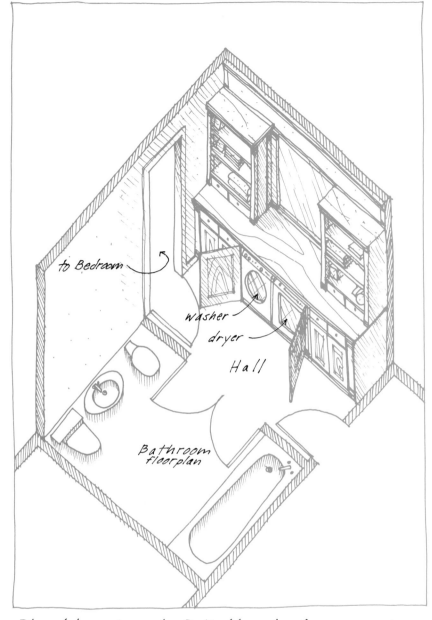

to Bedroom

washer
dryer

Hall

Bathroom
floorplan

Plan / Isometric at Bath / Laundry Arrangement

TV. An ironing board that folds into the wall will certainly save space, and if well designed will probably get used. Standard height for an ironing board is 31½ inches above the floor; this is for a "homemaker" of average height, 5 feet 4 inches. By all means, adjust the elevation of the board to suit your own requirements. In most circumstances, wall space is likely to be in great demand in the laundry room, and giving over precious real estate to a built-in ironing board may be impossible. You can always use a standard portable board and store it in a closet or corner.

Electrical outlets should be planned in convenient locations for ironing and other uses. If there's room enough, it wouldn't be a bad idea to provide a receptacle and cable hook-up for a small TV set.

If the utility bath area is to be located in the interior of the house, without either window or skylight, codes require a ventilation fan in the bathroom portion. An additional ventilator is also a good idea in the laundry area as well, because the air can get hot and stale.

Lighting

Designers like to divide the function of lighting into three categories: general, task, and decorative. What is needed in a utility bath is plentiful general and task lighting — which is to say an abundance of light. For daytime illumination, a window in this room would be useful, although in most utility baths the wall space will be commandeered by shelving or other forms of storage. If there is a roof immediately above the room, a skylight would provide excellent daytime illumination, but many budgets might not be able to accommodate the extra thousand or so dollars (installed cost) this feature will require.

Even with ample natural illumination, there should be plenty of artificial lighting for use at night and on cloudy days. The most common way to light such a room is with ceiling-mounted fluorescent fixtures. These produce a tremendous amount of light and are efficient to operate. In some jurisdictions, such as California, fluorescents may be required as an energy-conserving measure. Fluorescent fixtures have the additional benefit of giving off much less heat than incandescent lights, which is an important consideration in laundry areas, which tend to get overheated.

The cheapest and most common type of fluorescent lamp, called "cool white," produces a harsh, blue light that gives the skin a ghoulish cast and renders colors poorly. We prefer "warm white" fluorescent lamps, despite their extra cost, because of their better color rendition. But both cool white and warm white lamps distort one or the other end of the color spectrum, so with either of them you might have a little difficulty matching colors when you pull socks out of the dryer. You can alleviate this problem with what is called a "full spectrum" fluorescent lamp, which gives much more accurate color rendition. Full-spectrum fluorescent tubes cost more than warm white lamps and are not available in all sizes.

If money were no object, an ideal lighting plan might also call for recessed cans or ceiling-mounted incandescent fixtures directly above the appliances, sink, and work surfaces. This would assure ample task lighting and color balance on all these surfaces. But in a relatively small room, shadows will not be much of a problem, and it's a rare homeowner who is willing to devote so much effort and expense to lighting a utility room.

Flooring

Selecting your flooring material should also be governed by practical considerations. The floor should be decent looking, if not handsome, and easy to clean; but above all, it should be durable and waterproof.

Ceramic tile maintains its place as the top-rated flooring material in our minds, for all those qualities we have discussed: good looks, durability, ease of maintenance. However, I would be very tempted to use sheet vinyl throughout both utility room and bath as it is cheaper, somewhat easier to clean, and easier on the legs than tile. It makes no sense to spend money where it will not count, and this room is one of these cases.

In my current laundry room I have the original pine boards that have been in the house since it was built in 1836 — no sense changing them. But I would not install a wooden floor, just as I would not install carpet, stone, or any of the other flooring choices with the exception of flexible rubber flooring, which, though in the same cost category as the most expensive sheet vinyl, is handsome, easy to clean, and wears like iron. One testimony to the durability of rubber flooring is its use in the terminals at the Rome airport, one of the busiest in the world.

Drainage

An additional hazard faced by the utility bath floor is inundation — at some point, your washer may break down and send a flood of water onto the floor. In most utility rooms contained within the house, even those that are professionally designed and constructed, no provision is made for this eventuality; but it is poor practice.

One tactic, and the one we employed in the Dorchester project, is to slope the floor and install a drain at its low point. Washing the floor periodically should replenish the supply of water in the drain's P-trap, that sinuous run of pipe you see beneath all the sinks and lavatories in your house, which, by retaining a slight amount of water, keeps the sewer gases from escaping into the house. In Dorchester we put down ceramic tile. Since a tile floor typically is "water resistant," not waterproof, we installed a waterproof membrane beneath the flooring, much the way a waterproof pan is installed beneath a tile floor in a shower stall. The floor, with its slope and its waterproof membrane, turned out to be relatively expensive.

The more economical tactic, and one we've chosen for

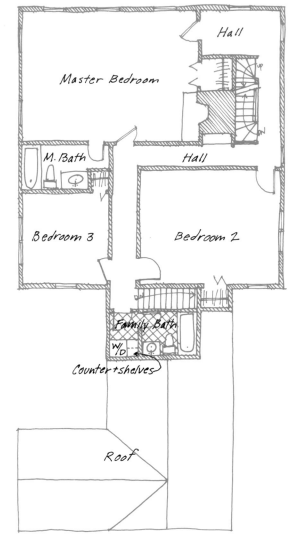

Utility Bath at Second Floor - Scheme One

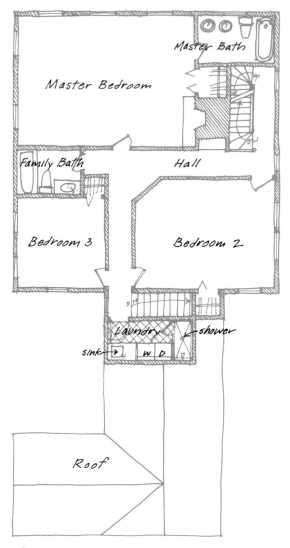

Utility Bath at Second Floor - Scheme Two

On the second floor we explored two differently priced options.

Scheme 1: Lower cost
Add a small master bath, renovate the existing family bath to include a utility area with stacking washer and dryer.

Scheme 2: Higher cost
Add both family bath and master bath. Renovate existing bath as laundry area.

nearly all of our other "This Old House" projects, is to place under the washing machine a high-lipped, galvanized metal pan, which is fitted with a drain. We recently used this device in our barn project in Concord, Massachusetts, in which we built a second-floor utility room adjoining a bath. Because no water would enter the drain in the bottom of the pan to replenish the water seal in the P-trap, our plumbing and heating expert, Richard Trethewey, installed a "trap primer," a small valve that releases a few drops of fresh water into the drain every time the toilet is flushed.

These preparations might seem like overkill, but I once lived on the top floor of an early 1800s mansion that had just been converted to condominiums. Although there were utility baths in each unit, the plumber had not installed overflow pans for the washing machines. Predictably, a washing machine (not mine) on the second floor leaked, and the resulting flood ruined the ceiling and woodwork of a lovely period room in the first-floor unit. The cost of the repair would have paid for the initial installation of drain pans several times over.

●

It might strike you that the type of utility suite we are describing here, with a full bath, laundry sink, storage, and so on, will be excessively costly. But consider that the functionality of the room will be determined by its size, design, and mechanical facilities, more than by the standard of the finishes. Spend your first money on design, reorganizing the space, and installing first-rate plumbing and mechanical systems. Save money by using sheet vinyl, high-pressure laminate counters, a fiberglass shower or tub/shower enclosure, and a stock lavatory and vanity, instead of tile, solid surfacing, and custom casework. There is nothing wrong with using plywood for the shelving and even the counters in the laundry area, provided it is well sealed with polyurethane. Remember, *utility* is what this room is all about.

Contemplating such a space might seem boring compared to dreaming about the luxury bathrooms we review later in the book, but this room is a workshop. The more efficiently you design and organize it, the faster you can get your work done and the sooner you can have a soak in that luxury bath in the master suite.

●

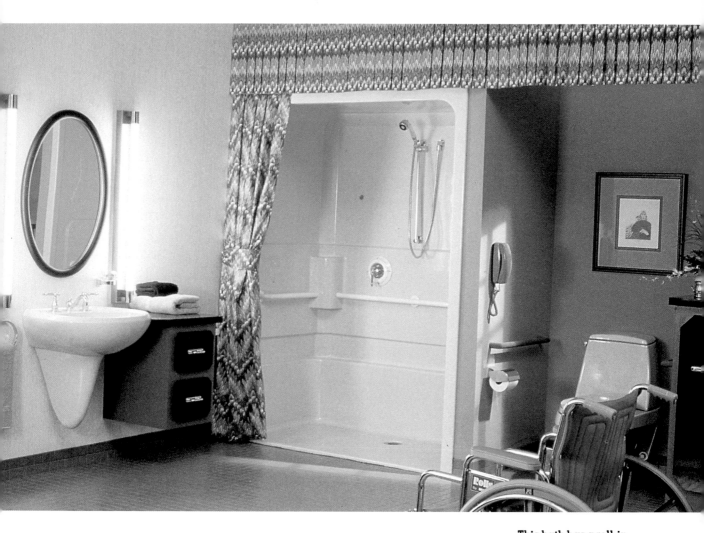

This bath has a roll-in acrylic shower stall, a wall-hung lavatory with integral protection of piping, a wall-hung toilet with transfer seat, and conveniently placed storage units. The phone in the bath is a good safety feature.

8 Barrier-Free Baths

The terms *barrier-free design* and *universal design* have mercifully replaced the dread rubric *handicapped accessible.* They are used to describe an architectural approach that utilizes standard appliances, materials, and construction techniques in designs flexible enough to be enjoyed by people of all ages and physical abilities — from children to the elderly, and from people with poor eyesight to those who must use crutches, a walker, or a wheelchair for mobility. At first glance there is nothing remarkable about some of the bathrooms pictured in this chapter, and that is precisely the goal of this design approach — to create bathrooms and living spaces that are appealing and functional for everyone. Anyone who's ever broken a leg knows how difficult it can be to get around the house. Doors become barriers instead of gateways to a new space, stairs become obstacles surmounted only with great trepidation and difficulty. The danger of slipping and falling increases dramatically, and just performing one's normal toilet is suddenly very clumsy and awkward. Having spent a significant portion of my reckless youth with various broken bones, I am sympathetic with this design approach. Among professional designers and architects, the approach is getting increased attention as older couples are renovating their homes with the idea of "aging in place" and as young families renovate in anticipation of taking care of an aging or disabled parent instead of placing them in a care facility. Finally, people with various levels of physical function are rightfully demanding that architecture take *them* into consideration.

Universal design seeks to create environments that respond to individual needs, building in flexibility to adapt to changing life circumstances and abilities so the occupants can function independently, safely, and comfortably. Most professionals working in the field strive to match the design of the particular room to the abilities of the person using it instead of designing to some theoretical standard. This design philosophy suggests the use of standard fixtures and construction techniques and, wherever possible, making

Special thanks to René Varrin, Paul Lasner of Specialized Housing, Inc. and George Welsh of Living Design for their help in preparing this chapter.

169

them accessible through minor changes in location, hardware, and mounting detail. This approach both keeps costs down and helps make the room look like a normal bathroom instead of a medical showroom. Overkill in design and equipment is not the goal; matching design and fixtures to the requirements of the intended users at reasonable cost and architectural impact is.

At "This Old House," we've experimented with barrier-free design. When we expanded a house in Lexington, Massachusetts, into a bed-and-breakfast, we made one of the bathrooms wheelchair accessible and looked at the design considerations for wheelchair access to the house itself. In our Concord, Massachusetts, barn rehab, architects Jock Gifford and Chris Dallmus arranged the floor plan to accommodate easily visits by the owner's father, who had increasing trouble climbing stairs. The first-floor study was made very open and accessible from the rest of the living spaces, but could be shut off with pocket doors to convert to a bedroom. The nearby powder room was sized as a full bath with a shower. In this chapter we've combined what we learned from our own work in barrier-free design with the experience of experts in the field. We hope that what follows represents practical, field-tested solutions to typical barrier-free design problems.

Location

The first issue in choosing a location for this bathroom is access. For the room to be reachable by people of many abilities, it should be on the ground floor. Once the person is in the house, he or she must have ready and independent access to the kitchen, living, bedroom, and bath areas.

Accordingly, in our concept house we converted the study into a bedroom and bumped out into the west courtyard for an adequate-size bath. We built ramps on the east side of the driveway to give wheelchair access. Note that ramps should be no steeper than one inch of vertical rise for every foot of horizontal run (1:12).

The second issue is the accessibility of the bathroom *within* the plan. Bathroom doors are often 2 feet 6 inches wide (called a "two-six door"), but when you subtract 3 inches for the door and hinges, the clear opening is only 2 feet 3 inches — too narrow for a person in a wheelchair to turn into from a standard 36-inch-wide hallway.

The literature recommends installing a 36-inch-wide door, but I am told that a door that allows a clear opening of 32 inches to 34 inches is usually sufficient. One of the problems with a 36-inch door is that it is hard for a person in a chair to open and close from the outside, as one's reach from a seated position is only about 24

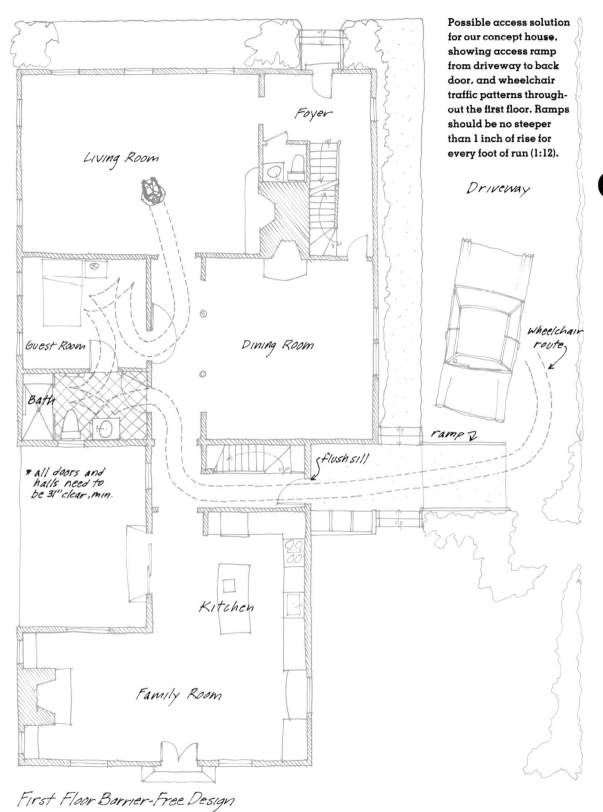

Possible access solution for our concept house, showing access ramp from driveway to back door, and wheelchair traffic patterns throughout the first floor. Ramps should be no steeper than 1 inch of rise for every foot of run (1:12).

Driveway

Foyer

Living Room

Guest Room

Bath

Dining Room

wheelchair route

ramp

flush sill

* all doors and halls need to be 31" clear, min.

Kitchen

Family Room

First Floor Barrier-Free Design

inches. (An inexpensive solution to this problem is to add a door pull or knob to the center of the door.)

Doors that swing into the room are hard to close, restrict the maneuvering room once inside, and are potentially dangerous, since anyone falling inside the room can prevent his or her own rescue. Doors that swing out are better. Another possible solution is a pocket door, provided it is engineered to be very easy to operate and does not have a floor-mounted track that would impede the wheels of a chair. A threshold as low as ¼ inch can stop a chair, forcing the driver to back up to take a run at it. Keep in mind that pocket doors can be problematic or even dangerous for people who have difficulty using their hands and arms, so in this and all design considerations you must match the design to the need.

Lever handles are more easily grasped and operated than knobs. A selection of very good-looking lever handles is on the market — so good-looking you may want to use them throughout the house.

Obviously, the considerations about door width, swing room at doorways, thresholds, and changes of elevation will be present throughout the barrier-free portion of the house.

Space Requirements within the Bath

The next consideration is to provide adequate area inside the bathroom. The average wheelchair needs a clear area 5 feet in diameter in order to make a full turn. There must also be adequate space in front of the lavatory, toilet, and bathtub or shower, and enough space between one fixture and another so that the person can transfer from the wheelchair onto the toilet or into the tub or shower.

Grab bars will be necessary next to the toilet and in the bath or shower area. Individual requirements should guide decisions about number and placement of grab bars. Generally, horizontal grab bars are used for pushing up, whereas vertical grab bars are used for pulling up.

A grab bar should have a nonslip texture and should be secured firmly to the wall studs so that it will not come loose even if grasped in the midst of a fall. You can give yourself some flexibility by sheathing the walls in ¾-inch-thick exterior plywood before the finish wall is applied. This way, grab bars can be securely installed anywhere and can easily be moved, as would be necessary if the bath was built for a growing child.

Experts recommend that towel bars be eliminated altogether in favor of grab bars, as any horizontal element will be grasped by someone in a fall and, clearly, a towel bar would not support the weight of a falling person.

A guest bath that doubles as a barrier-free bath. The room has ample (and nonskid) floor space for wheelchair-turning radius. Cabinet and lavatory height have been adjusted for use by someone in a chair.

Illumination

Natural light and ventilation are very desirable features in any bath, but the type and placement of windows must be carefully handled in a barrier-free application. A wheelchair user may be unable to open or close a double-hung window and may have difficulty reaching and adjusting the locking mechanism of a sliding window. Casement or awning windows easily operated by a crank mechanism are recommended.

The window should be located 24 to 32 inches above the floor. As in all bathrooms, avoid placing an operable window in a shower or above a tub. For people with vision problems, care must be taken to place the window so as not to create excessive glare in the room. (Also to control glare, designers caution against using dark colors on a window wall or using surfaces that are highly polished.)

In terms of artificial illumination, what's needed is ample light with a minimum of glare and shadows. As we age, we require more light — an octogenarian may need three times the light levels of a teenager. Moreover, some people with vision problems have trouble distinguishing walls from floors and the edges of counters from the surrounding roomscape. One technique barrier-free designers use is to place a stripe of contrasting color around the perimeter of the floor and at the edge of the counters to give the eye some reference point.

In terms of color, earth tones are to be avoided, because they absorb light. Most designers recommend using the brighter end of the color spectrum.

The need for high levels of illumination would suggest using fluorescent fixtures, which is just what we installed in the Lexington bed-and-breakfast project in the form of a "luminous ceiling" — fluorescent fixtures installed behind acrylic diffuser panels. But some people may have problems with the 60 Hz flicker associated with fluorescent lighting, so fluorescent fixtures should be shaded or baffled and used as indirect lighting sources. In general, indirect incandescent lighting is recommended for general illumination, combined with good incandescent task lighting in the form of pendant lights or recessed cans. Excellent illumination in the vanity area is imperative so that people taking medicines can easily read the labels on the bottles.

Light switches should be in a convenient spot, generally 30 to 36 inches above the floor. The switch itself should be whatever type the person finds easiest to use. Often a rocker switch, such as Leviton's Decora type, which responds to slight pressure, is easier to manipulate than a toggle switch, which requires more strength because one has to lift one's arm to use it and greater concentration because the toggle is a smaller target than the rocker panel. You may consider a delayed-action switch so that a person doesn't have to leave the room in the dark. Even better might be a switch controlled by a motion detector to turn the lights on automatically when someone enters the room and switch them off when they exit. If there are multiple entrances to the space, you will, of course, need three-way switches.

"Aging in place": This large bath was designed for a couple who anticipated retiring right where they were. The large ocean-view shower features a fold-down seat and grab bars; the natural-cleft slate floor has a good nonskid surface; there is ample natural light and room to maneuver.

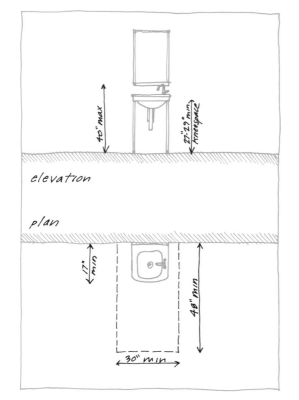

Lavatory Clearances
for Universal Bathrooms

Door and Floor Clearances
for Universal Bathrooms

Plumbing Fixtures

Vanities

The literature we reviewed on barrier-free baths recommends the use of a wall-hung lavatory to give a wheelchair user space to roll up to the lav with plenty of legroom. To maximize knee space, the P-trap can be placed near or even inside the wall, behind an access panel. All exposed supply and waste lines should be thermally insulated so the person does not burn his or her legs.

However, a wall-hung lavatory does not provide any storage for toiletries, so some designers use a standard lav with a cutaway vanity base. They rotate a round lavatory bowl 90 degrees to put a lever valve at the side, thus providing an accessible lavatory at the cost of a standard one. This can be coupled with an infrared faucet, such as one now sees in airport restrooms, which turns on automatically when one places one's hands in front of it.

The vanity height should not exceed 32 inches, with 29 inches of clearance below if the wheelchair has arms. Faucets should be no more than 12 inches from the edge of the vanity if they are to be within comfortable reach of someone seated, except where hand or arm coordination is a problem. Most designers prefer a sin-

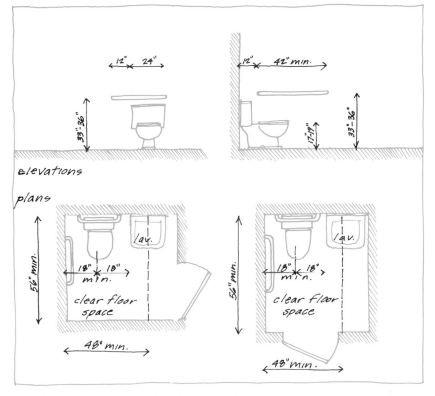

Toilet Clearances for Universal Bathrooms

gle-lever faucet or paddle knobs to twist knobs. Vanity tops should have rounded corners, and the mirror should be lowered or tilted to provide a full view of a person seated in a chair.

Adjustable vanities are available in both manual and electric models. The electric units are on the expensive side, but are an option.

No matter what lavatory is installed, electrical outlets (of the ground fault interrupter type), shelves, and cabinets should be installed within easy reach.

Toilets

If you're planning for a wheelchair user, the location of the toilet will depend on that person's technique for transferring from the wheelchair to toilet. Usually, the person will have a preference for transferring from the left or from the right, and this preference will affect both free space near the toilet and the placement of grab bars.

As in any bathroom, cleanup of the toilet and the floor area will be easier if the toilet is a wall-hung model, which might be mounted at the same height as the wheelchair to facilitate transfer.

After much experimentation, barrier-free designers George Welsh and Paul Lasner now specify a standard floor-mount toilet

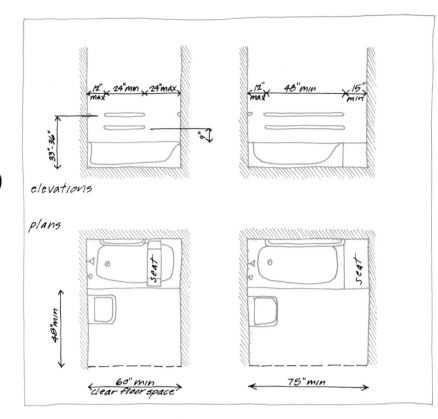

Bathtub Clearances for Universal Bathrooms

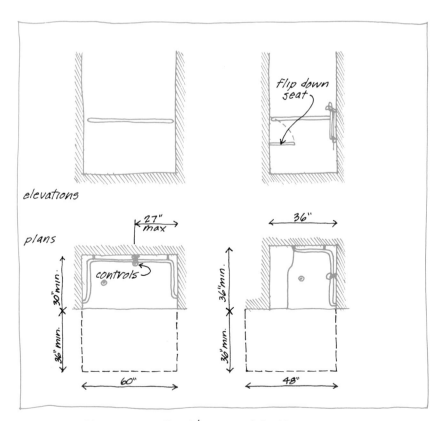

Shower Clearances for Universal Bathrooms

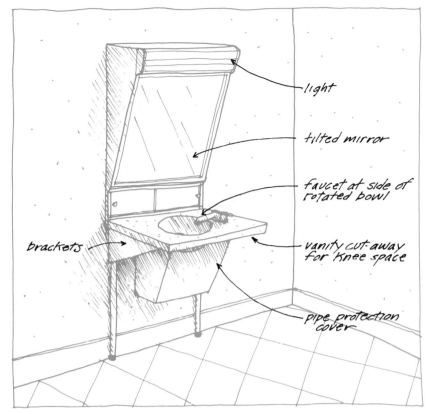

light

tilted mirror

faucet at side of rotated bowl

vanity cut-away for knee space

brackets

pipe protection cover

Accessible Lavatory

equipped with an elevated detachable seat. This option avoids any extra plumbing costs and structural considerations entailed in the installation of a wall-mount toilet, and also allows the greatest flexibility of use. Welsh further recommends installing a floor-to-ceiling grab post near the toilet.

Also available are height-adjustable toilets that can be equipped with a wash/drying jet option.

Bathing Facilities

The decision about what type of bathing facility to install is, of course, driven by the preferences and abilities of the person for whom the bath is being constructed and, to a certain extent, the rest of the family or caregivers. To keep general renovation costs down, one wants to build a bathroom that everyone in the family can use. So, while barrier-free shower enclosures are commercially available, they might not be suitable for a household with small children, unless the parents enjoy nightly water sports.

In most applications, a standard tub installed with some modifications is the most flexible fixture. The tub can be elevated on a platform to facilitate transfer from chair to tub, and a folding seat or transfer platform can be installed or built at one end. There

should be a minimum of 18 inches clearance between the outboard end of the transfer platform and any adjacent wall to accommodate the back of the wheelchair.

Horizontal grab bars should be installed on the wall not far above the top of the tub for support during transfer and bathing. Some people also like to have a stirrup grip suspended from the ceiling above the tub's outer rim. The stirrup grip should be secured to a joist with an eyebolt.

Bathtub controls must be easy to reach from both outside and inside the tub. Some people like side-mounted controls and a remote-controlled drain. A combination tub and shower requires a hand-held shower head (perhaps in addition to a wall-mounted shower head).

A pressure-balancing antiscald control valve is essential to keep the water temperature from turning hot or cold without warning. For those who cannot feel changes in temperature, a valve with a digital readout of the water temperature might be useful. For those with vision problems, an automatic tempering valve is available.

Shower Stalls
To avoid the tedium, hassle, and danger of transferring from wheel-

This bath represents the current state of the art in functional barrier-free design.

(Far left) Note the angled mirror, the open counter with insulated pipes allowing access for someone in a wheelchair, the single-lever faucet control, the lever door handle, and corner guards on walls.

(Left) The toilet has a raised seat and is flanked by a fold-down transfer seat and grab bars.

(Right) The tub also has a fold-down transfer seat, ample, well-positioned grab bars, adjustable hand-held shower, lever control kobs, and pressure-balanced antiscald valve. Both floor and walls are covered with seamless epoxy flooring.

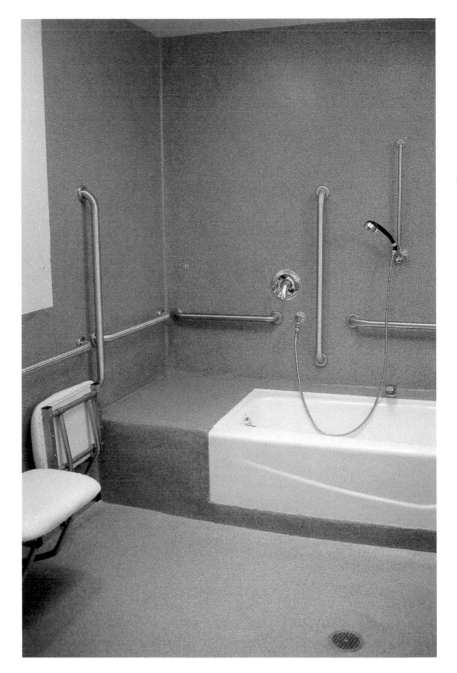

chair to bathtub, many people prefer roll-in showers. There are commercially available models with door tracks that are flush to the floor, which present no impediment to the wheels of a chair. Although a number of door styles are available, people working in the field say the fewer doors, the better, and they advocate a roll-in stall without a door.

One possibility is to locate the shower at the end of an L-shaped approach, with the walls far enough apart to accommodate the chair. In more restricted spaces, the shower can be part of a "wet-room" bath. The shower floor is sloped for drainage, but over-splash is allowed to go into the room. Clearly, this approach demands a waterproof room, with a very nonskid floor.

A roll-in shower should be 5 × 5 feet, although 4 feet square can work if space is tight. Such a shower, if handsomely designed, would avoid any look of a "handicapped device" that might deter a future buyer of the house.

It may be possible to make your existing bathroom accessible by installing a battery-powered lift system that mounts to a track on the ceiling. Such a system allows the person freedom of movement throughout the bathroom and to raise and lower herself or himself safely to or from a seated position in the bathtub, shower chair, or vanity. A voice-activated model may soon be available.

Shower controls must be placed readily at hand and equipped with antiscald valves. It may be desirable to be able to set the water temperature before entering the shower, in which case a digital readout valve placed some distance from the shower head may be called for.

Floors

As you would expect, the three requirements of floor-covering candidates for this bath are ease of maintenance, water resistance, and excellent traction when wet or dry.

Sheet vinyl is not a candidate because it is too slippery when wet. Ceramic tile is better, especially tiles with a nonskid texture. The grout lines give the floor some "tooth," too. But tile is still slippery when wet and must be carefully maintained, especially at the grout joints. It is also very hard, and a fall on a tile floor might result in greater injury than one onto some softer surface.

The preferred flooring material is a seamless epoxy floor of the type found in commercial kitchens. The material itself is a two-part mixture, a resin and a hardener that is applied over a mud-job or concrete substrate on the floor and marine-grade plywood or cementitious backer board on the walls. This forms a seamless floor and wall surface that is impervious to water. The coating can be colored with pigments and given texture with admixtures of quartz. It is generally installed by a specialist, but could be installed by an experienced amateur. One of the great advantages of this flooring material is that it can be used in a self-formed sanitary base or wainscoting, wherein the floor wraps up the wall, for a watertight, easily cleaned surface.

Another option that achieves the same effect at less cost is a polyurethane elastomeric, which is an applied coating to which you can add rubber granules to improve traction.

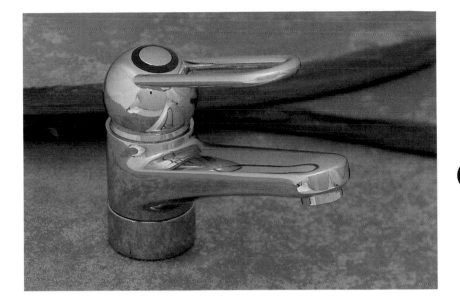

This single-lever faucet has two important safety features: a loop-handle that is easy to grab, and a scald guard control that allows presetting of maximum water temperatures. Great for a kids' bath, too.

Telephone

A telephone in this bath might be very handy, if not essential for safety. Also consider an intercom (the type that does not require pushing a "talk" button) and a panic button.

Construction Costs and Considerations

Construction costs for a barrier-free bath will almost certainly be greater than for a standard, economy bath renovation, if only because of the additional cost of extra lighting, wider doors, a custom vanity, and a custom shower enclosure. But by adapting standard fixtures as described above, and by building to a function instead of to an arbitrary standard, these extra costs can be kept to a minimum.

Since the ultimate functionality of the room will depend on the design and execution of many small details that might easily go unnoticed or that run contrary to standard building details, it is very useful to help the builder or carpenter see the world from the same perspective as the person who will use the space. To do this, a friend of mine who was building a barrier-free deck and exterior paths left a wheelchair on the site for the carpenters' and masons' use. Not surprisingly, they began to experiment with it, and whenever they wondered if a construction detail actually would work, they would hop in the chair and try it out for themselves. The problems addressed by barrier-free design and the necessity of rethinking the details became very real for them, and the whole crew took up the challenge of improving the job.

An unexpected result of both the design and the quality of

the construction is that the deck, patio, and walks are more pleasant for *everyone* to use. If you never saw a wheelchair on the premises, it would never occur to you that the house had been made "handicapped accessible."

This fact is not lost on the designers and architects I spoke to in preparation of this chapter, who believe that the barrier-free approach will start to influence architecture in general. The reason for this is simply because the feeling and functionality of our homes and commercial buildings are improved by incorporating some of the basic principles of barrier-free design.

●

Firms Specializing in Barrier-Free Design

National Rehabilitation Association
633 South Walnut St.
Alexandria, VA 22314

Specialized Housing Research and Development Corporation
14025 Farmington Rd., Suite 306
Beaverton, OR 97005
(503) 646-3582

Publications

Handbook for Design
The Department of Veterans Affairs (VA pamphlet 26-13)
Although published in April 1978 is still useful because of its clear, straightforward text and plentiful illustrations.

"Home Accessibility Information Series"
The Minnesota Housing Finance Agency
333 Sibley
St. Paul, MN 55101
Terse but filled with good information and sketches.

A Comprehensive Approach to Retrofitting Homes for a Lifetime
and
The Directory of Accessible Building Products
National Association of Home Builders
National Research Center
400 Prince George's Boulevard
Upper Marlboro, MD 20772-8731
(301) 249-4000

The Accessible Home
Specialized Housing Research and Development Corporation
14025 Farmington Rd., Suite 306
Beaverton, OR 97005
(503) 646-3582
Specialized Housing, Inc. and Living Design
Plans and specifications for families and institutional groups by a
nonprofit firm that works exclusively in the field of barrier-free design.

The Complete Guide to Barrier-Free Housing
Gary D. Branson
Betterway Publications, Inc.
P.O. Box 219
Crozet, VA 22932
(804) 823-5661
Covers barrier-free design considerations throughout the house. It is
well organized, clearly written, and well illustrated. It contains an
excellent bibliography.

●

9 Exercise Baths

This spacious shower room at the entrance to an outdoor pool gives *all* the bathers a place to shower and change. The Wesselmann nude, which is waterproof plastic, helps to set the mood for sunbathing.

Many of us have resigned ourselves to the fact that a regular aerobic and muscle-toning exercise program is necessary to look and feel our best. For me, at any rate, this exercise regime is not something I leap out of bed looking forward to each morning. It is a chore, and I do it only out of a sense of duty, necessity, and vanity. It is not surprising that many people have sought to design an exercise area into their homes, either because they can't wait to pump iron, or because, like me, the fewer barriers to performing the exercises, the more likely they are to get done.

An exercise room can be as simple as a cross-country ski machine stuck in the spare bedroom, or as elaborate as a small gymnasium, with weight machines, ballet barres, a steam room, and bathing facilities. Whatever you have the inclination (and budget) to build, remember that the exercise room is part workshop, which means it should be efficiently organized to allow you to accomplish what you need to do, and also part retreat, a place to escape from the quotidian in a feverish state of runner's bliss.

Location

A combination of exercise facilities and bathroom, here called the exercise bath, can be created just about anywhere in the house. A spare bedroom can readily be converted to this purpose, or, with a bit more work, so can the garage, attic, or basement — spaces that tend to be underutilized. The layout of your own house may suggest other possibilities. Whatever the potential location, it's generally desirable to give the exercise area visual and acoustic isolation. The sounds of someone pacing on a treadmill, dancing to aerobics cassettes, or lifting barbells is usually not music to other people's ears. For the peace of the household, the exercise bathroom should be located and constructed such that sound is not transmitted to the rest of the house. For your peace of mind, you should be able to do your workout in private.

●

The requirements of the exercise bath should reflect the particular exercises and equipment that make up your routine. Clearly, if you are a dancer, the room will be sited and organized quite differently

(Left) This spacious exercise room has plenty of natural light and ventilation and is simply furnished with a barre and an exercise machine.

(Right) This ultimate exercise/bath suite has everything — and a meticulously designed place to put it. The mirrored wall reflects a tiled exercise area with a mat, shelving for weights and towels, and a whirlpool with a hand-held shower. Walls and window shades are sound-muffling fabric. The TV/VCR can be viewed from the whirlpool or lounge chair. I'd be afraid to exercise here for fear of messing it up.

than if you pump iron. A critical requirement is the room's structural strength.

Architects and engineers describe structural loads in terms of "live" loads, that is, people moving about within the house, and "static," or "dead," loads, the weight of furniture, books, appliances, and so forth. An exercise area presents both: athletic equipment can add considerably to the static load that the floors and walls must support, and jumping or dancing, especially by two or more people in unison, can shake the structure with large live loads. We're not saying the house will fall down, but excessive flexing of the floor diaphragm can cause walls or ceilngs to crack, as well as create annoying reverberations throughout the house.

If you are contemplating a fully equipped exercise room in your house, we recommend you get a qualified architect or carpenter to analyze the existing structure and to specify any modifications to beef it up.

In an 1840s house "This Old House" expanded in Arlington, Massachusetts, we built an exercise bathroom containing a stationary bicycle and rowing machine, along with toilet facilities and a combination steam bath/shower. The exercise bath was on the second floor, above a new kitchen we were building at the same time. To carry the above-normal loads of the exercise area, we installed the floor joists 12 inches on center instead of the usual 16 inches on center. (The whole suite was designed to be converted to a drop-dead master bedroom–bath should any future owner so desire.)

These structural and environmental factors make the basement and garage prime candidates for siting an exercise bathroom.

Both live and dead loads are transmitted directly to the earth through the concrete slab; the space is removed from the living and sleeping areas, for visual privacy; and the earth absorbs much of the noise made while exercising, for acoustical privacy. In addition, the basement stays relatively cool throughout the year — better conditions for vigorous exercise.

Another structural consideration is the available headroom in the site of your exercise bath. Some activities, such as jumping rope, aerobics, dance, and yoga, and some weight-training machines may require a ceiling at least 8 feet or higher. A basement with a low ceiling can be given greater headroom by excavating the floor. Headroom in the attic can be extended by raising the roof or installing dormers. Surprisingly, excavating the basement or even opening one of the walls onto the yard may be less expensive and less disruptive to the ongoing life of the household than raising the roof in the attic.

Illumination

The room should have plentiful illumination. Windows or skylights may suffice during the day, but at night the room will need artificial lighting. In most instances, the best source is overhead fluorescent lighting because it produces plenty of light with little heat and is inexpensive to operate. Incandescent lighting is more flattering to the complexion and gives better color rendition, but does throw off a significant amount of heat, which may be undesirable in a room in which you are already sweating.

Climate Control

Ideally, you should be able to adjust the temperature of the exercise area independently from the rest of the house, as, in general, you want this space cooler than other rooms. An exhaust fan or some other mechanical ventilation would be useful, both to expel excess heat and to bring fresh air into the space. In many areas of the country, air conditioning is desirable, if not essential.

Outdoor Access

The ideal exercise bath would open onto a deck or patio, allowing you to move a bike, rowing machine, or NordicTrack outside to exercise in the fresh air. We have an expression here in New England: "The buds are on the trees, winter can't be far behind!" — winters are long, and in all but the very coldest weather, it is wonderful to get outside to exercise. It takes little imagination to envision an indoor-outdoor exercise area with sauna or hot tub and cold plunge reminiscent of Ten Thousand Waves.

A sunroom/exercise addition to this waterfront home allows the owner to enjoy the view from both steam room and massage table. Life can be tough.

Mirrors, Floor Covering, and Accessories

For many exercises, such as dance and aerobics, it's useful to have one wall, perhaps two, equipped with mirrors.

The most desirable floor coverings for an exercise area are resilient types, such as vinyl, rubber, and carpeting. Vinyl is cost effective and easily cleaned; rubber offers the same qualities at higher prices. Carpeting is probably the most versatile, offering even greater comfort, cushioning, and soundproofing while resisting damage from dropped weights and dragged exercise machines. Foam floor mats covered in vinyl are used by many people for aerobics and yoga.

For the bathroom area, we would of course avoid carpeting, preferring instead a simple, cost-effective, and easily cleaned surface such as sheet vinyl or ceramic tile.

Storage

The exercise area should contain storage for towels, athletic wear, weights, cassette tapes, and other articles needed for workouts and bathing.

While designing storage facilities, you might include a place for a television set and video cassette recorder for playing exercise tapes and for diversion during tedious exercises, as well as a stereo and radio for listening to music or catching up with the news on "All Things Considered." Don't forget places for a clock and a telephone.

(Left) An exercise space need be only as big as the rug you stand on to lift barbells and do some stretches.

(Right) A small carpeted room off the master bedroom combines exercise space with a dressing room.

General Space Requirements

The amount of space you need for your exercise area depends on what you intend to do in that space. Yoga or calisthenics which require no machinery might require a space of only 8 × 8 feet (although this is minimal), while a room large enough to accommodate a cross-country ski machine and a stationary bike might have to be 12 × 12 feet. The following are some general guidelines to exercise equipment and their accompanying accoutrements:

- A barre for ballet warm-up exercises is usually installed 1 inch above the user's hip bones; some barres are adjustable. Mirrors should be placed on the wall opposite the barre and at the barre's end.

- A cross-country ski machine (such as a NordicTrack) measures roughly 2 × 7 feet and requires an area of about 5 × 9 feet to operate. These machines are readily folded and stored when not in use, freeing up the same area for other activities.

- Stationary bicycles range from simple, fairly transportable models requiring perhaps 4 × 6 feet to larger stationary models, some of which are very sophisticated and can simulate the demands of riding through changing terrain. The larger models are not easily moved to make room for other activities.

- A jogging machine, which is a treadmill with grab bars on each side or in front, usually measures 5 to 6 feet long and 2½ to 3 feet wide. Most machines are not movable.

- Most rowing machines require a 6 × 3-foot floor area, but may be compact enough for closet or cabinet storage.
- Barbells require minimal space and should be accompanied by mirrors on the opposite wall.
- A weight-training center accommodates equipment for numerous strength-building exercises. A large free-standing exercise center may require an area as big as 15 feet square. A small, wall-mounted, single-station unit may need only 3 × 6 feet when in use and can be stored compactly.

A large, well-designed, and superbly equipped exercise room in a Long Island house. The bath area is behind the mirror. Doors open wide for fresh air and access to the terrace and waterfront views; the glass-block wall separates the bedroom from the workout area, but lets light filter through.

This whirlpool is located just off the greenhouse, where the steam rising from it can humidify the plants. A shower is on the platform to the right. Sliding doors seal the whole space from the adjoining bedroom.

The Bath

We said at the outset that the exercise bath is part workshop and part retreat, so the bathroom portion of the space might be as spartan as the utility bath, as practical as the family bath, or as luxurious as the large master bath.

When "This Old House" did our Santa Fe, New Mexico, project, I rented a large adobe house for my family. The master bath was, in fact, a large room with plenty of space for exercise equipment, and featured sliding doors leading to a private courtyard, double sinks, a toilet and bidet in a separate compartment, a steam bath, and a whirlpool. The room was great as a family recreation area, but lacked the requisite privacy for a master bath. As in all design decisions, the disposition of the room and the blend of luxury and utility you choose to factor in to your exercise bath project should reflect your needs, desires, and lifestyle.

Since I discuss these other types of bathrooms fully elsewhere in the book, I will restrict my comments here to general considerations and design comments pertaining to the large bathing facilities likely to be found in an exercise suite.

An outdoor shower is a way to wash kids on the way home from the beach. This cedar-paneled pavilion affords both privacy and an openness to nature. If it were mine, I'd shower in it all year (well, almost).

Extras

Saunas

Given the choice between a hot tub and a sauna (my experiences at Ten Thousand Waves notwithstanding), I would unhesitatingly choose a sauna. I find saunas to be more relaxing and, in a strange way, more social, perhaps because the well-established ritual of the sauna is a binding force between people. There is a practical reason for my preference as well — saunas require much less maintenance than hot tubs, which are in reality small swimming pools.

Essentially, a modern sauna is an insulated compartment that's heated to a temperature ranging between 160 and 195 degrees. A small gas or electric heater inside the compartment warms a tray of volcanic rocks, which in turn radiate their heat into the air. Humidity and, therefore, the apparent heat in the sauna are boosted by ladling water from the traditional wooden bucket onto the rocks.

The walls, ceiling, and slatted benches inside the sauna are generally constructed of a water-resistant softwood such as redwood or cedar. Both of these woods are good insulators and will not burn bare skin. The floor can be any water-resistant, easily cleaned material but is most commonly concrete covered with wooden slats, or duckboard. There may be a drain in the floor to carry away any excess water.

In my experience, you must follow the ritual to enjoy the sauna properly; it is not something you can truncate or in any way hurry. And, because you use the sauna naked, it is an activity done among friends. I learned the ritual from my friend Charlie Sincerbeaux, a Sybarite in stockbroker's clothing, with whom I share a passion for Oktoberfest and other rituals of endurance.

The sauna starts with a cleansing shower. Then, you enter the well-preheated compartment and ladle some water onto the hot rocks, which sends a gusher of steam into the air and produces a near-instantaneous sweat. You sit there for five to fifteen minutes, until your heart is racing and the blood pounds through your veins. Scandinavians further stimulate the circulation by beating themselves with a birch branch, an aspect not included in my training. When you've worked up a good sweat, you take a cold shower or, if you are lucky enough to have your sauna so situated, a roll in the snow or a plunge in a cold lake or pool. Then, after a short rest, you enter the sauna for a second session, followed by a cold plunge if you want to be reinvigorated or a lukewarm shower if you want to stay soothed and mellow. There follows a long cool-down, during which, clad in only a towel, you and your fellows drink the requisite aquavit. Thus relaxed and transported, you are able to share your views of life from the proper perspective.

If this is the way you intend to use your sauna — and many of these remarks pertain to hot tubs or spas, as well — then clearly you will need to create a space in which you and your guests are encouraged to detach yourselves from the world. Some facility for dressing is required, equipped with a bench, cubbies for jewelry, watches, and glasses, and a closet or pegs for clothing. You should have storage for large towels as well. Showering and toilet facilities should be nearby, preferably in the same area as the sauna, so that

A traditional Finnish log sauna on a Maine farm. When I came across these photos in a magazine several years ago, I fell in love and became determined to build one of my own someday.

once you enter that intimate area, you do not have to go back into public space until you are ready to face the world again. The tub or plunge, if you have one, should be nearby, as should the resting area. Ideally, this would be a place outdoors where you could view the softly lit garden and the night sky, or perhaps in a room enclosed in sliding doors, which could be opened in the warm months and closed in the cold.

In keeping with the atmosphere of detachment, a telephone or television is not desirable (in my opinion), whereas a stereo for playing ethereal music is.

My dream sauna, one I hope I can build someday, is very much like the saunas still built in Finland. It is a freestanding struc-

ture in the garden, situated near the sea for cold plunging. It would be down a path and away from the house and office, so that from the moment you set off, you are detaching yourself from the world. Such a sauna could be heated by a wood stove — the firebox is located on the outside of the structure — which aficionados claim produces the finest quality sauna heat. Obviously, to build such a sauna requires the proper location, which at Boston real estate prices requires winning the lottery.

But fortunately, complete saunas are available in kit form, the assembly of which would make a good do-it-yourself project. You can also build your own from scratch and outfit it with commercially available temperature controls, electric sauna heater, and other hardware. As we mentioned above, cedar and redwood are the preferred species for the paneling and benches because of their water and rot resistance. However, sauna builders warn against using aromatic cedar of the type used in cedar closets, because when heated to 160 degrees or more, it reportedly gives off a vapor that causes difficulties in breathing. Inexpensive woods such as knotty pine and spruce are not stable enough to stand up to the heat and dryness of the sauna.

One-person saunas are manufactured as small as 3 × 4 feet, which means they can fit into some closets. (My friend Charlie had managed to wedge one into the master bathroom of his Boston Back Bay condo.) A common size for a family sauna is 5 × 7 feet. Remember, bigger is not necessarily better. The larger the space, the more time and BTUs will be required to bring it up to the temperature. Ideally, there should be one wall at least 6 feet long to enable you to stretch out. The larger saunas have benches on two levels, the top level being hotter. The ceiling should be low to conserve heat — 6½ to 7 feet is common, and walls and ceiling should be very well insulated.

The sauna should have soft lighting suitable for meditation or perhaps reading. Use a surface mount instead of a recessed fixture to avoid cutting holes in the insulated walls, and make certain it is approved for moist areas. As in any location that brings water and electricity together, all electrical circuits should be controlled by a ground fault interrupter, even if local codes do not require it. The heater should be activated by a sixty-minute timer, which will shut the unit down when the time runs out.

Every sauna should have a means of bringing in fresh air for the bathers. Usually there is a vent, but the same purpose can be achieved by undercutting the door at floor level. Commercial saunas must also have an exhaust vent to ensure constant replenishment of the air, and although some residential saunas do not have this feature, it is recommended. Standard specifications require a vent area that equals 5 percent of the floor square footage or 1½ square feet, whichever is greater.

For safety's sake, there should be no locking mechanism on the door. The door should open outward so that a bather who collapses inside does not impede his or her own rescue.

Steam Baths

Steam baths, like saunas, stimulate one's circulation by elevating the body temperature. The difference between the two methods is that steam baths use very moist heat instead of the sauna's dry heat. Once steam baths were only found in locker rooms at the YMCA, but in recent years compact steam generators have been developed

A modular steambath
can fit into most bath-
rooms.

for residential use. A standard tub or tub/shower enclosed to retain
the steam is used as the "steam room," making the residential steam
bath an option for many bathrooms and budgets.

    The steam bath cycle is similar to that of the sauna. You sit
or stand in the shower compartment for ten to twenty minutes while
temperatures climb to 110 to 130 degrees. You can take your cooling
shower right there, which causes the steam to condense; then, repeat the steam cycle. At the end of the steam session, you again
shower to cool the body and close the pores; then, relax and cool
down thoroughly before getting dressed.

    Steam baths can also be used for temporary relief from re-
spiratory problems, such as colds, sinus conditions, allergies, and

asthma. A special steam head can mix eucalyptus oil or other me-
dicinal substances into the steam, making breathing easier.

     In comparison to building a sauna, installing a steam bath
is quite straightforward. As we have discussed, the compartment itself is a standard shower or tub enclosure slightly modified to entrap
the steam, which is produced by a small steam generator. Operating on 120 or 240 volts AC, the generator is fairly compact and could
be installed in an adjoining closet or vanity, or in the attic above or
the basement below the bathroom. Water from the domestic hot
water supply is plumbed to the generator, which converts it to
steam. The steam is delivered by another pipe to the steam head inside the shower compartment. To control the system, a timer and
thermostat are installed outside the shower compartment. We recommend the installation of an additional air switch inside the compartment to enable you to control the system without stepping out of
the shower.

**An exercise solarium
opens onto the garden in
this town house. The dou-
ble sinks were placed
here to create more room
in the small adjoining
bath, which houses a toi-
let and shower. An un-
usual arrangement, but
as the solarium connects
to the master bedroom, it
all works well as an intimate space. Excavation
made the ground floor ac-
cessible to the outside, so
one can have plenty of
fresh air and light during
exercise. The green-
house in the garden
houses an art studio.**

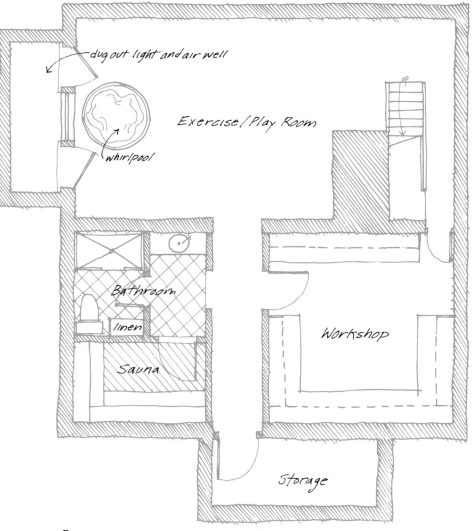

dug out light and air well

Exercise/Play Room

whirlpool

up

Bathroom

linen

Sauna

Workshop

Storage

Basement Exercise Room/Bathroom

Coverting the basement can often free up additional space at minimal cost. This plan gets natural light and ventilation into the exercise area via a light well. Given sufficient space in the side yard, one could create a small patio in the excavated space and minimize the feeling of being in a ditch by terracing up to ground level. A basement exercise area can be used without fear of disturbing sleeping children or spouse.

It is essential to seal the shower compartment, or else the whole bath will become the steam chamber. Hinged shower doors equipped with continuous magnetic seals do a good job, as do sliding doors equipped with felt or rubber gaskets. Unlike a conventional shower, a steam shower compartment has no opening above the door. In commercial steam baths, which operate continuously throughout the day, the ceiling is sloped so that condensing steam does not drip on the bather. In a residential steam bath, however, the slope is often omitted to simplify construction.

Most steam bath compartments are surfaced in ceramic tile. Andrew J. Shabet, of ThermaSol Steambath Company in Ridgefield, New Jersey, suggests using a grout hardener when setting the tile, to make the grout tougher under the onslaught of heat and moisture. He also suggests the grout be treated with a silicone

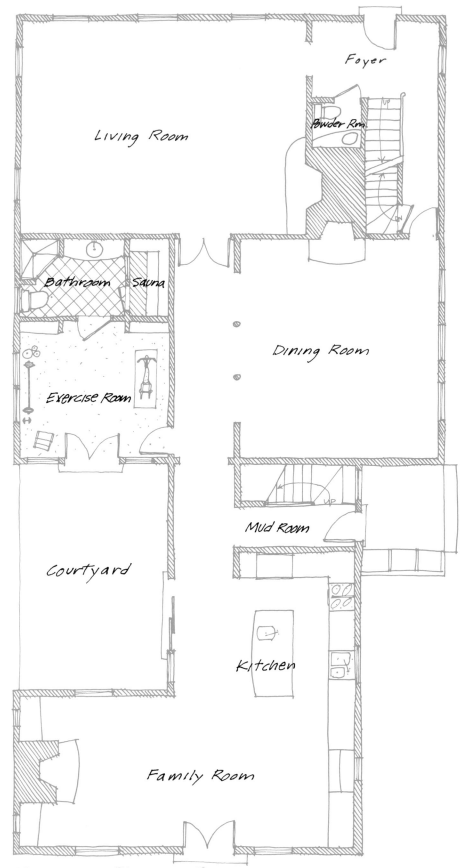

204

Living Room

Foyer

Powder Rm.

UP

Bathroom   Sauna

Dining Room

Exercise Room

Courtyard

Mud Room

UP

Kitchen

Family Room

First Floor Plan with Exercise Room and Bath

We explored a number of options for the exercise bath in our concept house, creating additional floor space by commandeering the entire study and by pushing the back wall out into the courtyard.

Scheme 1:
Separate sauna, toilet and shower/lavatory.

Scheme 2:
Sauna opens to bath. Two-person shower.

Scheme 3:
Whirlpool overlooking exercise area and outdoors.

Scheme 4:
Whirlpool with corner shower.

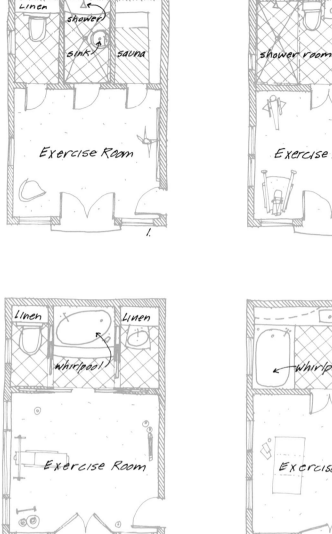

Exercise Rooms / Baths

sealer. One disadvantage to tile is that it is a good conductor of heat, which means that when the steam is first turned on, the cold tile will cause the steam to condense until the walls warm up.

Good alternatives to ceramic tile are acrylic or fiberglass paneling or solid surfacing, all materials that are better insulators than tile and therefore cause less initial condensation, enabling the steam to accumulate more quickly in the chamber at start-up.

A steam bath can be retrofitted to an existing shower enclosure. The procedure is much the same as for new work, requiring the installation of the steam generator, plumbing lines, steam head, and controls. If the bathtub or shower already has doors on it, they would have to be modified to prevent steam from leaking around their edges. The compartment itself must be sealed at the top to prevent steam from escaping, either by building a soffit to lower the compartment ceiling or by installing a vertical glass or plastic panel above the shower doors to confine the steam. Installing or retrofitting a steam bath is not really a job for the average do-it-yourselfer, as it requires the expertise of a plumber and an electrician.

You should be aware that steam baths require regular maintenance. Shabet says that the water strainer in the steam generator should be cleaned every three to six months to remove mineral deposits. The generator's inner tank will also collect mineral deposits and should be cleaned at least three times a year, more frequently if local water conditions demand it. If not regularly used, Shabet recommends that the steam generator be run for about ten minutes every three weeks to prevent buildup of mineral deposits.

The cost of operating the generator is remarkably low. A twenty-minute steam bath may consume 2.2 kilowatt-hours of electricity — at a cost of less than twenty-five cents in most localities. In that time, you would have used less than a gallon of water.

Jetted Tubs

There are two general classes of whirlpool baths, or "jetted tubs": the whirlpool bathtub, which is filled and drained with each use; and larger spas and hot tubs, in which the water is kept hot and ready for use at all times. In the latter class, there are several types.

"Portable spas" are manufactured acrylic or fiberglass tub units with integral pump and filter, and a plastic or wood surrounding skirt. These units are generally shallow enough to fit through a standard doorway and are installed simply by placing them on a smooth level surface, such as a deck or patio. Some units operate on 110 volts AC and plug directly into an outlet, others must be hardwired to a dedicated 220-volt circuit. The units are essentially "ready for use" upon leaving the factory.

A second type is a manufactured acrylic or fiberglass tub which is built in to its destined bath or patio setting. The mechanicals may be in the same enclosure as the tub, or removed from it in a separate housing.

The third type is a custom or semicustom spa or hot tub of concrete or wood, built from custom-fabricated and standard components.

Whole books have been written about the design, installation, and care of spas and hot tubs and it is senseless here to try to duplicate that information. (Sunset Books has a good primer on the subject.) But in dreaming of and designing your exercise bath, master bath, or ultimate bath, consider the following points.

Whirlpools

The whirlpool bath involves a quantum jump in expense above that of a regular tub, and like any large appliance, installing one requires thorough planning. Floor joists may have to be bolstered or an extra hot water heater installed. In addition, you will have yet another machine to maintain (the pump and lines in the whirlpool require *monthly* cleaning) and repair when it breaks.

The whirlpool bath's massaging effect is produced by the stimulating effect of the moving water and the air bubbles bursting against the skin. This increases blood circulation, which soothes sore muscles. So the design of the jets influences the particular therapeutic effect of the whirlpool. Whirlpool tub manufacturers follow one of two different approaches to jet design. One employs a few jets with large outlets to give comfortable bathing, as softly pulsating water swirls around the body. The other approach makes use of more smaller-diameter jets to deliver higher-velocity streams of water. The latter design may not produce as comfortable an overall hydromassage, but may be more effective at soothing specific muscles or areas of the body.

Another point to note is that some jets, usually on less expensive tubs, are constructed to inject air into the top of the stream of water. The better jets, by contrast, inject air into the center of the water stream, where it more thoroughly mixes with the water, resulting in a superior massage. Also, pay attention to the material the jets are made of; the best are chrome-plated brass, of lesser quality are plated plastic. The most flexible jet systems allow you to control the spray direction, volume, and the ratio of air to water by adjusting an air switch mounted on the tub itself.

While whirlpool tubs may look the same, there is considerable variation in quality, a difference usually reflected in the price of the tub. The major plumbing-fixture manufacturers such as Kohler, American Standard, and Jacuzzi produce jetted tubs that satisfy requirements of the Underwriters Laboratories, the American National Standards Institute, and other quality-assurance panels that govern the industry. To get a tub that has gone through rigorous testing procedures, you naturally must pay a higher price than for some of the jetted tubs produced by smaller companies that don't submit their products to such thorough testing and regulation.

There are hundreds of what are called "jetters" — little companies that buy tubs and then add jets, pumps, and controls to make them function as whirlpools. Their projects are usually less expensive. Some of them work well; some do not. Unless you can thoroughly investigate the manufacturer, there is some risk that the tub bought from a jetter may not function properly. If the tubing that dis-

recessed installation, enameled cast iron
42" x 60" x 15"

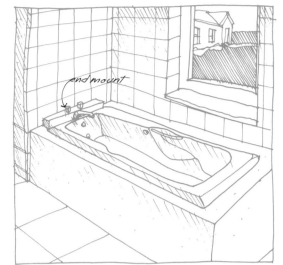

grab bars, slip resistant surface
36" x 72" x 18"

high gloss acrylic finish
34" x 66" x 18"

retrofits normal tub location
32" x 60" x 16"

fiberglas reinforced polyester
44" x 70" x 18"

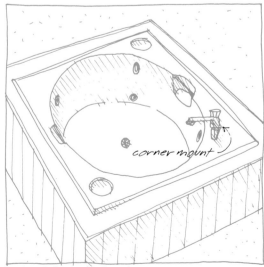

adjustable whirlpool jets
72" x 60" x 21"

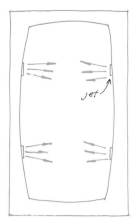

Specific Point

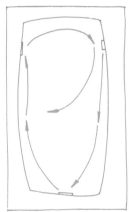

Fish Hook

Whirlpool Jet Placement

(Above) It is the air bubbles in the water that massage the skin. A soft massage or an invigorating one is determined by the amount of air induced to the water stream. Less-expensive systems inject air through a port above the water stream, where the bubbles rise to the surface and never burst against the bather's skin. A better system injects air into the center of the stream for a more complete air/water mixture.

(Left) Whirlpools come in all sizes and shapes, with a variety of amenities. Some styles are available as soaking tubs, without pump and jets.

tributes the water around the tub is not pitched accurately, for example, water will remain in the tubing when the tub is drained. Bacteria can then grow in the warm water, which will contaminate the next bath. If the jetter uses inadequately supported flexible tubing, it can sag from the heat and weight of the water, also creating pockets of bacteria-laden water. The best tubs use rigid copper or plastic tubing throughout.

Plumbing expert Richard Trethewey says that when problems arise, it may be much more difficult to get a satisfactory settlement from a jetter than from a major manufacturer. Some jetters buy tubs made by major manufacturers, but this typically is of little help to the consumer if the whirlpool bath malfunctions. The Kohler Company warns that when its tubs are outfitted by a jetter, their warranty may be void. At "This Old House," we have learned through experience that, in plumbing, as in all aspects of building and renovation, it is cheaper to install a high-quality product at the start than to have to repair a cheaper but inferior product down the line.

A good whirlpool is equipped with controls that can be operated while the bather is in the tub, without fear of electrical shock. Nobody wants to have to get out of the tub to adjust controls installed on a wall several feet away. The unit should be equipped with a built-in timer to shut the pump off if inadvertently left on.

In a well-designed whirlpool, the pump and all major components will be placed so that they can be serviced from one end of the tub. It is advisable to engineer into the tub surround some sort of access door or hatch that can be opened or removed without damaging the bathroom's finished surfaces. In my master suite, the access door to the whirlpool is beneath a window-seat banquette in the adjoining bedroom. Since at some point you may have to overhaul the machinery, design the whole tub surround so that it can, if necessary, be removed. Be sure to buy extra tiles and store them with the pump so that any damaged tiles can be replaced. Tile colors vary slightly from batch to batch, so the extras should be from the same batch installed during the remodel. Store a schematic of the tub surround with the literature from the tub so that you can take it apart with minimum damage.

Whirlpool baths come in a variety of sizes and shapes. Some of them feature reclining backrests, contoured bottoms, armrests, and built-in seats. They range in cost from hundreds of dollars for a standard-size tub fitted with jets, to tens of thousands of dollars for American Standard's whirlpool extravaganza, the "Sensorium," equipped with "Ambiance" — that is, a microprocessor temperature-control system for the tub along with a TV, intercom, picture

Exercise / Utility Rooms and Baths

phone, and video surveillance system that allows you to see who is at the front door and even unlock it remotely.

Spas and Hot Tubs

Whirlpool baths are generally installed in a standard — if perhaps luxurious — bathroom setting, within a master suite or perhaps in an exercise bath. Spas and hot tubs, in contrast, are generally installed in a room or outdoor space created especially for them. The remarks I made concerning the sauna and its requisite dressing areas, showers, toilet facilities, and general aesthetic considerations apply to spas and hot tubs as well. The experience of escape and re-creation that comes from soaking in a tub is partly physiological but mainly psychological. As in fine dining, the setting, the presentation, the company, and the sense of ritual are key.

If you are dreaming of a hot tub, you probably envision it being outdoors. If so, remember that provision must be made for screening it for privacy and from unwanted views, and for protecting it from wind, insects, and accidental entry by pets or small children. Since the tub and surrounding features will make a large design statement in your yard or outdoor area, you should think of it in terms of the overall site and landscape master plan. Once you install the tub, you don't want to have to move it to make room for walks, driveways, or plantings or for repairs to sewer, water, gas, or buried electrical or telephone cable.

Three different plans incorporating the requirements for laundry, exercise, and family bath, all on the first floor.

Sybaritic exercise room with motivational model in the corner.

The other thing to point out while you are still dreaming of your hot tub is that it will require a good deal of maintenance. One of the things that made the experience at Ten Thousand Waves so pleasant was that at the end of the evening we could put on our clothes and go home. Someone else would have to monitor and adjust the water chemistry, clean the filter and skimmer, inspect and fix the pumps, and periodically drain, scrub, rinse, and refill the pool. All these tasks must be performed daily, monthly, or seasonally. Some people enjoy such tasks and would gladly trade the time it takes to do them for the experience of soaking in the tub while gazing at the stars. Others (I include myself in this category) have trouble enough finding time to mow the lawn and keep up with routine house maintenance. Before the design and building program for

This built-in whirlpool anchors one end of a magnificent poolhouse. Imagine bathing in front of the fire, the January snows swirling outside the windows. For all the money spent here, though, I would have contrived to have the whirlpool communicate directly outdoors through french doors.

your spa or hot tub takes on a life of its own, ask yourself to what extent it really fits into your budget and lifestyle.

The point here is not to discourage anyone from building and using a hot tub or spa. Quite the contrary. One of the things most American homes lack is a place for true relaxation, meditation, and renewal, a place to retreat from the world. By carefully evaluating how you work, relax, and entertain, you can design and build an area that you and your family and friends will use. For my friend Charlie, it was his sauna. For you, it might be a hot tub, a lap pool, or, for that matter, a Japanese tea house or simple garden pavilion. You want a space that is an extension of your personality so that both you and your guests will feel, as we did at Ten Thousand Waves, privileged, expansive, and at home.

●

VOLUME 23

10 The Master Suite

Cool California modern. A glass tub is the focal point in this innovative suite of rooms, where the architectural emphasis is on unusual materials and informal arrangement.

It is axiomatic that good planning will help any renovation project proceed more smoothly and produce a better final result. But this principle is especially important in design-intensive areas of the house like kitchens and bathrooms. As we've said before, the majority of bath remodels will take place within the confines of the existing room, so the larger issues of space planning and design rarely become part of the discussion. Not so when it comes to the master suite.

I think of this space as an interlinked set of four discrete areas: bedroom, bathroom, dressing, and sitting. Other spaces, such as the kitchen and the utility bath, are also interlinked, but the master suite is distinguished from them all by the fact it is the least public, most intimate area of the house. Normally, a visitor (one with any manners, anyway) does not think of entering this space without an explicit invitation. The master suite is the inner sanctum, the last bastion against the outside world. So has it been since medieval times, when the very ownership of a bed was a privilege of the aristocracy, and the "privy-chamber" was the most secure room in the castle.

The ideal master suite would receive ample natural light for reading the Sunday newspaper while drinking morning coffee (and other daytime activities) yet would be very private both day and night. It might have space for a TV set or perhaps a sound system and would contain a sitting area separate from the bed, where one could snuggle up with a book without disturbing a sleeping spouse or partner. It might have access to a private outdoor area, a small deck or balcony, where one could ponder some last philosophic musings before turning in for the night, or invigorate the spirit with fresh air before charging out to take the world by storm in the morning. On the functional side, the suite should contain ample storage for clothing and room for dressing, and must be possessed of a well-planned, well-executed bathroom where one can let the mind expand while the body relaxes. Finally, the overall disposition of these spaces should be such that one entering the suite will pass through successive layers or zones, each more intimate and private than the preceding one, until reaching the "privy-chamber" of the sleeping area.

Such a master suite sounds luxurious and therefore expensive. But remember that we are talking about the arrangement of space and not the nature and cost of the materials finishing that space. Even if you cannot have a private balcony, a whirlpool tub, and marble bath in your master suite, you may well be able to achieve a space that functions well while giving you a sense of privacy, serenity, and intimacy. Good planning and good design are the key.

If you already have a well-designed master suite and merely need to refurbish the bathroom itself, you are fortunate. We cover the master bath specifically in the next chapter, but before you begin to think about what tiles, fixtures, and accessories you want for your new bath, pause to consider the master suite as an ensemble. Designing this area is first a space-planning problem, then an interior-design problem, and finally a materials and mechanical-engineering problem. You may find some simple ways to make your existing master bedroom much more useful.

Looking for floor space for their master suite, the owners of this house expanded into an adjoining bedroom, creating both character and separation with columns and bookcases.

Space for this master
bath was carved out of
the existing bedroom.
The frosted-glass parti-
tion allows light through.
The bath is compact yet
gets it all in: double pedestal sinks, whirlpool,
dressing table, and win-
dow seat. The WC and
shower are in a separate
room behind the camera.

Finding the Space

Here in New England a good deal of the housing stock was built
long before the advent of central plumbing, so it is not surprising to
find no architectural accommodation for bathrooms in the original
floor plans. In the 1940s, many old houses were retrofitted with a
family bath on the second floor and perhaps a half-bath under the
main staircase on the first. The idea of a full bathroom hidden away
in the master bedroom area must have seemed an impossible extravagance.

Even in new construction, up until about thirty years ago,
most houses were equipped only with a family bath and perhaps a
half-bath. A master bath was found solely in houses for the prosper-
ous. Now, however, it is virtually standard in new construction, for
reasons that are easy to comprehend — it offers adults more privacy, it eliminates family conflicts over use of a single bathroom,
and it's a lot handier to get to in the middle of the night.

A master bathroom is equally desirable in an older home,
although often it is neither easy nor cheap to install, considering the

A curved cherry wall encloses the bath. Outside the curve is the lavatory and master dressing area.

difficulty of carving out space for the bath or master suite (which may involve structural modifications to the building), running plumbing and electrical lines from the basement to (typically) the second floor and plumbing vents to the roof, and installing fixtures and finishes. While a master suite, or a newly renovated or installed master bathroom, might be an attractive selling point when you put your home on the market and may increase the ultimate price, in most cases you will not fully recover your investment. If you plan to sell your house in three to eight years, our general advice would be to do the minimum in the master bath or suite. Look for cost-effective ways to improve the functionality of the space: change traffic patterns by adding or eliminating doors or free-standing closet units, improve the light and views with windows or skylights, or simply refurbish the decor. Many times, what is wrong with bathrooms is not

A sensual, Art Deco re-
treat that promises mo-
mentary escape into the
delicious world of a by-
gone era.

so much the basic layout of the space, but an accumulation of small details that make the room look or function poorly: a stained sink, leaky or outdated faucets, mildew growing on the tile grout due to poor ventilation, inadequate storage, lighting, and mirrors, a battered vanity. Do not underestimate the impact of new fixtures, paint, wallpaper, and decoration.

Especially in an old house, bathroom renovation can be an all-or-nothing proposition. My wife and I once owned a lovely Greek Revival house built in 1846 and last plumbed probably in the 1940s. The supply pipes were a mixture of galvanized steel, copper, and threaded brass (mostly the latter), and the waste lines were cast iron and lead. The master bathroom had been installed in a sizable bedroom, the most dramatic feature of which was the large fireplace (*every* room of the house had a fireplace). Since the bathroom was far larger than necessary, we explored the feasibility of repositioning it, thereby gaining an additional bedroom. But doing so would have required reconfiguring a sizable portion of the second floor, and since we did not plan to be in the house for more than three years, we rejected this course. The next level of renovation was to replace all the plumbing fixtures, but that would have involved re-plumbing most of the house. Old brass piping, like a wild beast, is usually harmless if left alone. If you do try to modify it, the resulting shaking and wrenching on the pipe can break unseen connections buried in the walls. In the end, we left all the fixtures — the claw-foot tub, the antique toilet located in its own closet with a small window overlooking the garden, and the pedestal sink — and merely decorated around them. When we sold the place, one of the features the buyers especially loved was the antique bathroom with the fireplace.

●

If a master bathroom or suite is worth building, it's worth giving it space enough to function well. A cramped master bathroom hardly justifies the trouble and expense to build it, since you will likely be frustrated the moment you and your spouse try to use it at the same time.

My own master bath is 6 × 12 feet, dimensions I consider right at the minimum for a bath that *feels* like a master bath. That gives enough room for a moderate whirlpool tub (3½ × 5 feet), a separate shower stall, a toilet, and a single lavatory. The room would feel much smaller without the skylight, large windows above the tub, and mirror across one whole wall. The room works fine, although a single sink is a serious drawback. If I had it to do over again, I would have extended the room three feet in the long dimension in order to have two lavatories.

(Top left) The master bath, at 7½ × 12 feet, is at the lower limits of "master proportions" — the large windows, skylight, full-wall mirror, and counter extending over the toilet and tub make it seem larger than it is. Next time I'll make room for double sinks. I incorporated a toe kick at the base of the bath pedestal, which helps one reach across or into the whirlpool.

(Top right) In my own master suite, bed and bath look northwest over the backyard. Although in the city, trees screen our views in summer and the setting spring sun shines through the windows. In all seasons, my wife loves to contemplate her garden from the window seat.

(Bottom) The room behind the bedroom was intended to be a "master den," but it never quite worked out. We plan to install freestanding closet kiosks to provide dressing areas for both of us.

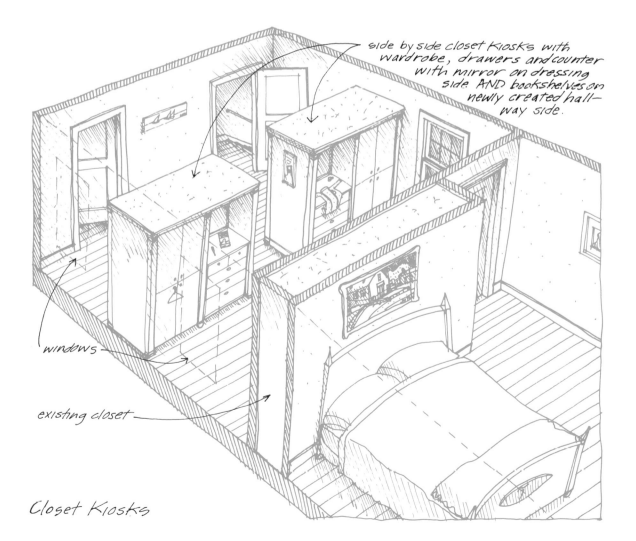

side by side closet kiosks with wardrobe, drawers and counter with mirror on dressing side. AND bookshelves on newly created hall—way side.

windows

existing closet

Closet Kiosks

Enclosing the shower and tub in a ventilated glass surround enabled a dressing area to be incorporated into this narrow city bathroom. The curved wall in the foreground encloses the stairway.

The point here is that space is one of the hallmarks of a master bath. The room ought to be big enough to let you bathe, shower, shave, or put on makeup without feeling jammed in. The extra room is partly a matter of aesthetics but also a matter of necessity; this is the only bathroom in the house likely to be simultaneously used by two people.

Master Baths in a Limited Space

If you are forced by budget or structural limitations to build the master bath in a small area, consider maintaining a sense of spaciousness by installing fewer fixtures: a shower and no tub, or a tub and twin lavatories with the toilet placed in a separate compartment.

A master suite carved out of the attic packs a lot of functionality: small work area to the right behind the bed; walk-in closet to the left, leading to the skylit bath. The headboard doubles as a dresser. I particularly like the feeling of being up and away from the rest of the world.

Even though tubs, toilets, and lavatories are all available in down-sized models, we would discourage you from using them for fear of cheapening the sense of the room. The tub shouldn't be a shrunken model that makes you curl your knees up out of the water; nor should the shower stall be so tight you can hardly bend to wash your legs without bumping into the walls.

What types of fixtures you need in your master bath are determined to an extent by your habits. If you and your spouse never use the room at the same time, perhaps you do not need two lavatories. If you take only showers, perhaps you have no need for a tub. However, beware of constructing a room that is so idiosyncratic it will discourage a future buyer.

Another strategy in dealing with inadequate space in which to build a full master bath is to divide functions between more than one bathroom. Typically, men prefer showers; women love at least an occasional soak in the bath. So one approach might be to place a large tub in the main master bath, and a more spartan shower stall in a bath down the hall, which the man could use when guests were not in the house.

●

Logically there are only three ways to provide the additional space a master bathroom or suite requires. First is to expand horizontally into existing rooms; second is to expand horizontally by adding on; and third is to convert an underutilized portion of the house, such as an attic or even a basement. In terms of cost, expanding horizontally into existing space is usually the cheapest, followed by converting attic or basement space. Adding to the footprint of the building is generally, though not universally, the most expensive option.

Master Suites

shower/toilet

Bath

Closet

Bedroom

Sitting

1.

One of the alternatives for getting a master suite into our concept house was to build out over the kitchen and family room below. (We explore other alternatives in chapter 1.) The resulting L-shaped space gave us good south-facing orientation and views of the garden — natural advantages that could be exploited in a number of ways. Six plans explore the possibilities of this space. Each scheme reflects our general philosophy that the bedroom area should be the farthest removed from the common areas and that bed and bath should be as visually and acoustically separated as possible, using the dressing room for the buffer. It is not always possible to do this, but we think it's desirable.

Scheme 1:
A simple arrangement with modest bath and dressing facilities but large sitting and bedroom areas.

Scheme 2:
A little more unified bath here. I would flip the bedroom and sitting room so the closet did not open onto the bed.

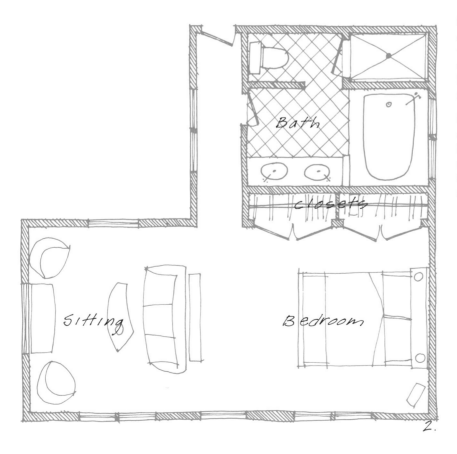

Bath

closets

Sitting

Bedroom

2.

Scheme 3:
Bigger bath and dressing area at the expense of space for the sitting area. Sitting area in this plan could be used for a small home office.

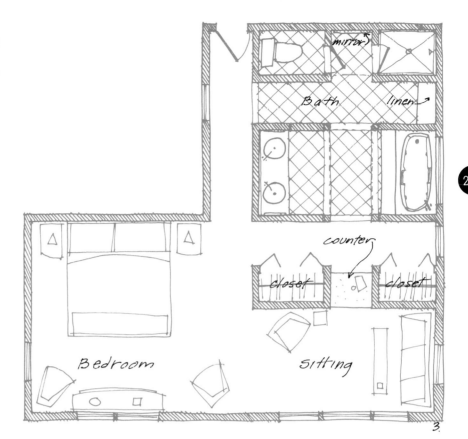

One of the best sources of horizontal expansion space is an unneeded bedroom, especially in a big old house that's blessed with many. We used this tactic in the very first "This Old House" project, in which we built the master bath in what had been a small second-floor bedroom. The house already had three reasonably large ones, so sacrificing the fourth, which was cramped anyway, was an easy decision.

Other targets for horizontal expansion might be extra or easily replaceable closets, staircases, or covered porches. I found space for a master bedroom suite in my own house by eliminating a back staircase and gutting a small apartment that had been inserted in the house in the 1940s.

In the Santa Barbara craftsman-style bungalow project of "This Old House," the only space for a master suite was in the attic. But to convert it to living space, the roof had to be raised, the existing ceiling joists had to be reinforced to carry the new loads, and a new stairway had to be squeezed into the already-tight floor plan. What we ended up with was a spacious, 15 × 28-foot bedroom, with a bathroom, dressing area, and plenty of storage.

The difficulties in expanding upward, though real, should not be exaggerated. Plenty of old houses can accommodate a top-

Scheme 4:
Dressing area placed behind the headboard. This scheme masses sleeping, dressing, and bathing on one side, with sitting on the other, which may be desirable if one spends a lot of time lounging in the bedroom.

floor master suite with little structural modification. Such was the case in the master suite "This Old House" built beneath the gables of an old frame house in Melrose, Massachusetts. The roof was tall enough not to need raising, the existing stairway was adequate, and the sloping walls of the attic added character to the room. The resulting space contained a full bathroom, bedroom, and even patio doors leading out to a small balcony. The general point here is: don't be deterred by possible structural problems until you know for sure what it will cost to surmount them.

When space for a master bathroom cannot be found inside the house, the only remaining choice is to build an addition. Usually, this is the most costly solution, involving a new foundation, walls, and roof. However, in the case that extensive structural modifications must be made to, say, an attic space to enable it to carry the additional loads posed by a master suite, adding on might actually be cheaper. In return for its cost, an addition often gives you the maximum freedom of design: a new bathroom or master suite hobbled by the fewest compromises with existing conditions. You might even use such an addition to solve other design flaws the house has, such as lack of a mudroom.

Scheme 5:
This scheme is probably the most practical for a two-career family where both partners have to get up and dressed in the morning and spend their late evenings in bed (falling asleep over their legal briefs) rather than lounging in a chair. Two banks of closets, plenty of access to dressing area and bath, and a pleasingly straightforward layout.

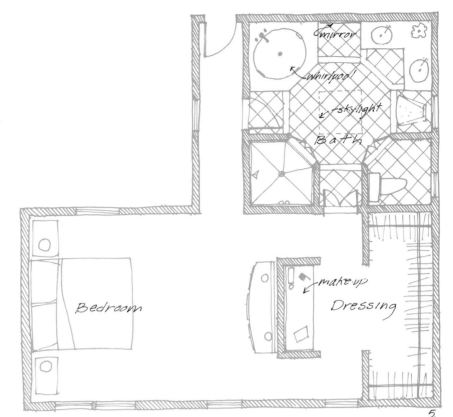

Scheme 6:
A variation of 5 with some intriguing possibilities for unusual room geometries and selected views of the garden through windows in the makeup area. If I were building a master suite in this space, I would look seriously at scheme 5 while exploring permutations of 6.

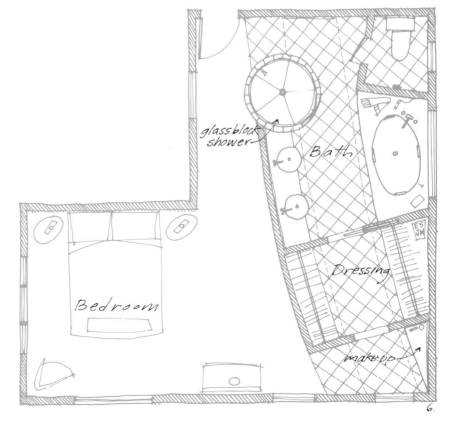

Meshing the Bathroom with the Bedroom

The master bathroom needs to have a "comfortable" relationship with the master bedroom. Exactly what that means will vary, depending on your way of living. In the course of doing "This Old House" projects in California, New Mexico, and Louisiana, one approach we've seen is to merge the master bath with the bedroom, with only partial screening between bed and bath. Sometimes there is even a direct line-of-sight from bathtub to bed. This, of course, heightens the sense of spaciousness and plays up the idea of romance, as if every night held the promise of an extended honeymoon. But "real life" (as it is called) usually involves getting up in the morning, showering, shaving or applying makeup, dressing for work, and heading out the door. Usually, you want some privacy for these mundane ablutions and do not wish to be on stage, even if the only member of the audience is your spouse or partner.

If personal privacy is important, so is restricting to the bathroom wet towels, shaving cream, cosmetics, hair dryers, toothbrushes, deodorant, dentures, and other paraphernalia instead of having them spill into your sleeping or sitting area. Nor is it desira-

Since the master bedroom suite is *your* private retreat, give it your own style. This one has the feel of a ladies' sitting room in an English country house. The sink is fitted into an antique marble-topped dressing table.

Small "hers"-style bath efficient enough to double as a dressing room. Sometimes separate bathrooms are the best solution for the master suite — no more whiskers in the sink, right, ladies?

ble to let the cooler, dryer air of the bedroom mix with the warm, moist atmosphere of the bathroom, thereby making the bath chilly or the bedroom muggy. Finally, there is the issue of noise. Unless both partners are on the same schedule, one of you will be disturbed by the other's routine bathroom functions.

The consensus at "This Old House" is that an open-plan bathroom might be great in a Las Vegas hotel or at a beach resort, but for day-to-day functionality, it has too many strikes against it.

Protecting the Inner Sanctum

For approximately eight hours of every twenty-four — more, if one spouse works nights — the bedroom is given over to sleeping. One should be able to use the bathroom without disturbing the other person's rest.

The placement of the bathroom door makes the most dramatic difference. In the typical arrangement, the door opens onto the sleeping area; thus shielding the bathroom from sightlines outside the bedroom. Orienting the door away from the sleeping quarters might keep light, foot traffic, and to an extent noise from disturbing someone in bed.

There is no "right" way to position the door linking bedroom to bath. It's a matter of weighing what's most important to you and arriving at a solution in keeping with your style of living. A new bathroom in an older house rarely turns out perfectly; concessions must usually be made to the shape or orientation of the space available. These limitations aren't entirely bad; sometimes they give the bathroom character. But as you do your analysis, remember that the key consideration is to preserve the bedroom's privacy and sense of intimacy.

A second issue is noise. The activities of one's daily toilet involve showering, flushing the WC, running the water in the sink, singing, and so on, all sounds that would disturb most sleepers. The noise is especially noticeable if the bathroom fixtures and the bed are on opposite sides of the same wall, just inches apart. To a certain extent, the generation of noise can be reduced by the types of fixtures you install — some models of toilets, for instance, are quieter flushing than others — but in general, noise must be controlled by the design and placement of the bath itself.

The transmission of sound can be more actively controlled by constructing a sound insulated wall, by placing the noisiest bathroom fixtures away from the bed wall, or by situating the bath so that a closet is placed between the two rooms.

(Left) A proper gentleman's dressing area wherein the suits speak for themselves. Jeeves, have you seen my bowler?

(Top right) Teddy Roosevelt's dressing room and bath at Sagamore Hills is both stately and comfortable.

(Bottom right) "Elegant rusticity" in the master bath of this gentleman's ranch. The retro-look tub and hardware, lots of counter space, and a big, well-lit mirror gives the room the saloonlike feel of a Wild West hotel.

A comfortable master bath/dressing area shields the bedroom at left from the hallway at right. Ample floor space and extensive use of ceramics give the room a hacienda feeling.

Placement of the Dressing Area

Another mistake I made when I did my own master suite was that I didn't sufficiently think through the dressing area. Not only is this space essential for the storage and donning of clothing, but its placement and arrangement offers the designer another way to protect the inner sanctum of the sleeping area by positioning the dressing and closet spaces nearest to the hall or foyer leading into the suite. Again, you want to feel that you pass through successive zones, each more intimate and secluded than the last. Also, putting the dressing area near the entrance eliminates the need for closets in the sleeping area, which both gives it a sense of specific purpose — sleeping and lounging — and frees up wall space for windows and window seats, sitting areas, or other furnishings.

As a sweeping generality, the dressing room is, for a man, the most utilitarian area of the suite and, for a woman, an area of great intimacy. It almost goes without saying that the areas for the two partners should be separated — lack of definition of dressing fiefdoms is probably one of the things that motivated you to do the renovation to begin with. As in the master bath, these areas should be arranged to suit your lifestyles. For couples who like to dress in privacy and silence, complete segregation of the dressing areas might be in order. For couples who like to discuss the events of the upcoming day or social evening, integration of the areas might be just fine.

In any case, we feel that the dressing areas should be used as part of the buffer for the sleeping area. Ideally — and in all the plans we developed for our concept house — place a small entry

Another ladies' sitting room featuring a nineteenth-century French bureau with marble top, an antique American wicker chair, and a cross-stitch rug. The custom marble sink top echoes the dresser.

A masculine dressing room/bath. No expense was spared on the mahogany casework or marble. The tub is flanked by a shower and steam room through the frosted-glass doors.

area or foyer at the door leading into the master suite to lend a sense of transition. This is followed by the master bath, then a dressing area, and finally the sleeping area. But as in any design problem, there are no hard-and-fast rules; every design, especially in renovation, will be a compromise among lifestyles, budget, and factors specific to your site: light, views, the shape and size of the available space.

Outfitting the Closet

With respect to the design of the closet space itself, here again, one's lifestyle and personal preferences come into play. Before I became host of "This Old House," I owned, in addition to jeans and T-shirts, a total of two dress shirts, three ties, and two sport coats — plus a white linen suit I bought on impulse in Hong Kong from a tailor named Sam. This sartorial abundance hardly demanded organized closet space, which is part of the reason I didn't pay any attention to it during the renovation. (Now that I've added two — count 'em,

two — suits to the inventory, it is becoming an issue.)

In most households, however, both partners have a variety of work and leisure garments for all seasons that must be organized and stored so that they are well kept and easily retrieved. Some people prefer everything to be on display and easy to get at, while others want their clothing discreetly hidden behind doors, creating closets with a tidier, more formal appearance. Some want a bank of built-ins; others prefer individual pieces of furniture. Built-in storage utilizes space more efficiently and can make it easier to store and retrieve items, whereas furniture can lend the dressing room an Old World flair.

Until a decade or so ago, there was no question about how to outfit the closet — you installed a horizontal pole for hangers, with a shelf above it, and that was that. The bigger the closet, the longer the run of poles.

Poles are still an efficient and simple means of storage, although the closet pole has been superseded to an extent by the

Two well-appointed but not extravagant dressing rooms. Both are carved out of adjoining bedroom space, and are large enough to keep everyday articles close at hand and to store out-of-season clothing and sports gear. When my wife saw the photos, she looked at me meaningfully and said: "*This* is what I want."

closet "systems" that are widely available. One rule of thumb has it that 6 feet of hanging storage should be allocated to a man and 12 feet to a woman; but like all rules of thumb, this one may not necessarily meet your needs. The most foolproof way to determine your storage requirements is to measure your wardrobe to see how much you really need.

If an architect is designing your master suite, he or she will no doubt suggest the overall shape and size for the closet and dressing area, together with some rudimentary ideas about how to outfit it. But to then have the architect actually design the closet may be a waste of his or her time and your money. For that kind of detail, you may want to turn to an interior designer, a bathroom designer, or one of the growing number of closet specialists. Perhaps the most direct way to tackle the problem is to formulate the design yourself, using the ideas that these specialists employ.

A good place to start is by calculating how much space you need for full-length garments. These will need the hanging space provided by a traditional pole. But many items, such as shirts, blouses, suits, and slacks, can hang in much less vertical space,

which allows you to condense your closet to two levels of hanging storage: a total of eighty-four inches vertical height is considered adequate. Ideally, the pole heights should be adjustable, since wardrobes and styles change as time goes by.

Drawers are great for storing sweaters and all sorts of items, but not if you forget what's in them. Excessively deep drawers provide excellent hiding places at the backs. Some designers suggest limiting the depth of drawers to 14 to 16 inches, so that the sweater your mother-in-law knitted will not remain at the back of the drawer until it is too moth-eaten to be worn.

Another technique is to install drawers with clear plastic fronts, or use wire-grid drawers so that the contents can be identified at a glance. Drawers can also be outfitted with divided trays, for consolidating collections of small items like cuff links or jewelry.

Shelves are an excellent alternative to drawers. A variety of sizes and spacings can accommodate the range of garments to be stored. Shoes can go on a sloped shelf with a lip on the front some distance above the floor, eliminating the need to get down on all fours to find your favorites.

Wire-grid shelving is increasingly popular for closets and laundry rooms. Manufacturers like to call it "ventilated shelving," as the open grid lets air circulate through the clothes. Wire shelving is light, simple to install, and economical, especially compared to the cost of priming and painting wood shelves, in addition to the cost of the stock. It is a good project for a do-it-yourselfer.

Closet Maid and other companies manufacture complete closet systems, most of which are available at home centers and lumberyards. The components of the systems include wire-grid shelving, shoe racks, hanging racks, hooks, and sliding bins that take the place of drawers and are great for socks, underwear, and lingerie. Bins make things visible and easy to retrieve and allow the air to circulate through the clothing.

Although the system may appear to be expensive, one must factor in to the equation the ease of one-stop shopping (all the components for a full and functioning closet are available in one place) and the fact that anyone the least bit handy can install them as a DIY project. Then too, the closet systems can be changed to meet the needs of, say, a growing child or teenager.

Solid shelving, the major alternative to wire shelving, comes in a wide choice of materials: wood, fiberboard, and particleboard covered with laminate, vinyl, or melamine. In general, these composite materials are stable and fairly cost effective. Standards, brackets, and the shelves themselves are available in most home centers and lumberyards. The shelving comes precut in various lengths: 2, 4, and 6 feet. Laminate-covered shelving is more expensive than the wire grid, but it does have a more polished appearance. Like wire grid, it is prefinished, so no further work is required on your part. We would avoid using unfinished composite board shelves, because they are hard to clean.

The cedar closet is the traditional place to store off-season clothing. Cedar has an aroma pleasing to humans but not to moths, which makes it an ideal material for lining closets. Full-thickness cedar is fairly pricy these days, but you can get the look, the aroma, and the benefit of the species by using thin cedar paneling. Milled out of solid stock, this product consists of tongue-and-groove boards about 3/16 inch thick and bundled in packages of random lengths. It is smooth on one side and rough on the other, so you can choose the look you want. It is easily and accurately cut with a hand or power miter box and can be glued to a wall with wallboard adhesive. Making a cedar closet by lining an existing one with this material would be an excellent do-it-yourself project.

Lighting and Mirrors

Natural light is undesirable in a closet because ultraviolet radiation

Compartmentalized drawers for jewelry, lingerie, and shoes can be hidden behind doors.

(Left) Freestanding cedar closets in this New York City loft double as a room divider.

(Right) A master suite that evades preconceived notions of "bath." The most expensive element here is the custombuilt counter; the rest of the bath is comprised of simple materials. Imagination and a good eye for design are far more important than budget.

can fade clothing. Wherever possible, artificial lighting should be used exclusively. Generally, incandescent lighting is preferred because it provides more accurate color rendering than fluorescent lighting. Recessed incandescent ceiling fixtures, which place the light source up and out of the way, are a good choice in a closet or storage area, but if your ceilings are high enough, an economical porcelain fixture that accommodates a bare bulb works fine too. Homeowners with especially generous budgets might consider low-voltage lighting, which offers very accurate color rendition.

In the vicinity of the closet or in a nearby dressing area, there should be a long mirror, situated so that you can stand several feet away and get a full view of yourself.

●

Whatever the nature of your house and your concept for your master bath and suite, become very involved in the design process. Not only will you likely spend more money renovating this area than the other baths in the house, but the master suite will be *your* intimate domain. When it is finished, you want to feel comfortable, secure, and "at home."

●

The Master Bath

First-class design and materials come together in this bath: cherry vanity and tub surround, travertine marble counter, tub top, floor, and wainscoting. Note the separate room for toilet and shower and the towel warmer. The coved ceiling echoes the ellipse of the vanity, and both walls and ceiling are finished in glossy paint — tricks that make the room feel larger than it is. The same design could have been well expressed in less costly materials.

Developing a Floor Plan

If you are renovating or adding an entire master suite, then you've probably been compelled to do some planning of the master bath as well. The number and placement of the lavatories, the size and placement of the windows, tub, shower, toilet, and bidet all have an impact on the size and shape of the bath's floor plan and thus the corresponding bite it takes out of the rest of the suite. If you are renovating just the master bath (or have skipped the last chapter), it is still worthwhile to think these issues through, as a properly designed and executed master bath must dovetail with the master bedroom it serves. In the last chapter we tackled the general space-planning issues. Here, we need to focus on the particulars of the master bath itself.

Lavatories and Makeup Areas

Unless you are installing a large spa or whirlpool that will become the focus of attention, the master lavatory and vanity are likely to be the largest mass in the room, with the greatest run of horizontal surface to command the eye. The lav is also the bath's most frequently used element, so engineering the right elevation and location, with all the necessary accoutrements close at hand — soaps, makeup, shaving cream, towels, etc. — is essential. (The lack of such a work area is probably one of the things that motivated the renovation in the first place.)

Common sense and standard practice suggest the lavatory be set into a counter with storage beneath. Drawers are generally more useful than the open space filled with pipes — so the more drawers, the better. You can tailor their sizes and shapes to fit your needs and habits. Frequently used toilet articles such as electric shavers, hair dryers, and contact lens steamers should be within easy reach. You might even contrive a drawer equipped with an outlet to enable you to leave these appliances plugged in.

You probably will want double sinks in your master bath, one for you, the other for your spouse. In laying them out, make sure there is enough room beside each bowl for toiletries and grooming equipment. The National Kitchen and Bath Association says 11

inches between the bowls is the minimum; 18 to 24 inches would be better. At the end of the counter, allow at least 6 inches between a bowl and the wall — 12 inches if you can — so you don't feel jammed into a corner when using it. It's no fun to bump your elbow against the wall while trying to shave.

Some couples like to have two lavatories side by side so that they can organize their day while performing their morning ablutions. Others prefer solitude. Although lavatories are customarily placed side by side to utilize the same supply and waste pipes, nothing prevents placement on different walls. Some designers situate one lavatory where it commands a view of the outdoors. The placement of lavatories should take such habits and preferences into account — again with the caveat not to do anything so idiosyncratic that it would deter a buyer of the house later on.

Here is a balanced handling of space with an eye to practical materials. Double sinks, a solid-surfacing countertop, excellent lighting at the mirror, ample storage, and a tub with a view are all set off with a band of hand-painted tile.

Twin pedestal sinks and medicine cabinets make a delightful focal point in a dressed-up old-style bath.

If space is tight (as it was in my own master bath), consider installing just one lavatory. It is better to have one large, well-situated bowl with ample room around it than two small ones crammed into an inadequate area. If two people rarely wash up at the same time, eliminating one bowl may, in practice, pose no hardship. Decisions like this force you to be clear about your habits and priorities.

Another possibility is to install a separate vanity just outside the bathroom, typically in the woman's dressing area. Many first-class hotels use this tactic to provide a place for making up in a more spacious and tranquil situation while freeing up the bath for one's partner. Upon reflection, it is indeed a curious custom we have of building but a single space in which to excrete, bathe, and beautify ourselves. The economics that initially drove the design of the modular bathroom no longer apply. There is no reason — especially in a master bath — to be enslaved to an old design paradigm.

The standard height for the top of the lavatory counter is 30 to 32 inches, but professional designers generally suggest raising it to around 36 inches — standard height for kitchen counters — as many people find this elevation more comfortable for reaching and bending. Any counter space at which one will sit, such as a makeup vanity, should either be tabletop height (27 to 29 inches) or be furnished with a stool.

244

A confidently designed and executed retro-style bath with plenty of details worth noting. At the sink: marble over-mounted on lavatory, twin recessed medicine cabinets, good mirrors, makeup mirror, ventilation, and lighting. In the bath area one has a choice among bath, hand-held shower, or big shower head that gushes a broad, even sheet of water.

In a custom-designed bathroom you can tailor all these dimensions to your particular ergonomics. You can even place each lavatory at a different height, yours to suit you, your partner's to suit him or her. A five-foot-two woman will have different preferences than a six-foot-three man. The key is to be able to use the lavatory without excessive bending. Experiment by placing a washbasin with a mirror in front of it at different heights to see what is most comfortable.

A mirror directly in front of the lav is not always necessary; one placed on the side wall or even on a retractable arm might do the job just fine. In general, women need more mirror area than do men, and prefer an arrangement whereby they can see their faces from several different angles. Three-part mirrors work well, as do mirrors on swing-out arms.

Lighting

Lighting around the mirrors and lavatories is of special importance. There is nothing worse than giving yourself a good look-over in the

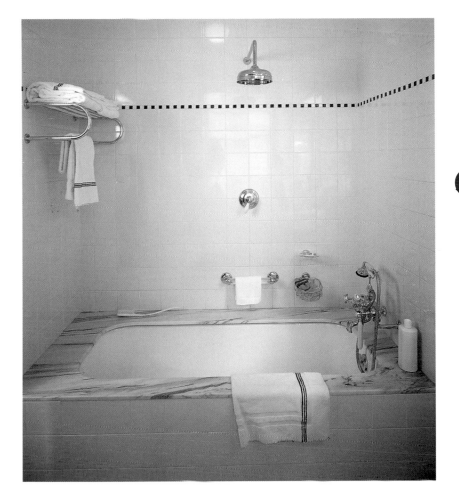

morning only to conclude (however secretly) that you look like hell.
If only for purely psychological reasons, choose lighting that is flattering to the complexion. For this reason, fluorescent lighting, despite the fact it is efficient to operate and throws off plenty of light, is
to be banished from the master bathroom. Even the so-called
"warm white" fluorescent lamps will make just about any complexion look like one of the habitués of a *Star Wars* bar.

For shaving and applying makeup, theatrical lighting in the
form of rows of exposed, low-wattage incandescent bulbs is still the
system of choice. Theatrical fixtures are readily available, in many
lengths, shapes, colors, and finishes. For even illumination that
avoids shadows on the face, chin, and neck, run fixtures down the
sides of the mirror as well as along the top.

If you are planning a master bath with a whirlpool or soaking tub, consider installing dimmer-controlled, recessed down lights
above the tub and vanity (in addition to makeup lighting). At full
power, the down lights will provide general illumination for moving
about the room, when cleaning, or when searching for your false

teeth. Dimmed down, they will create pools of soft light, perfect mood lighting for romantic soaks in the tub when the moon is full.

If the toilet is in a separate compartment or otherwise segregated from the bath, you will need to provide illumination (and ventilation). The simplest and least intrusive might be a recessed down light. The same holds true for the shower. Most recessed down lights are rated for use in damp (but not wet) locations. These are the most utilitarian areas of the bathroom, so the key thought is to select fixtures that will provide good illumination without making an overpowering design statement.

Finally, do not overlook the possibility of shared illumination, both by day and at night. One of the most appealing features of the third-floor suite "This Old House" did in Melrose, Massachusetts, was a wall of glass block between the bedroom and the bath, which gave the whole room a sense of light and space, without making it feel exposed.

Showers, Bathtubs, and Whirlpools

All of us at "This Old House" consider a walk-in shower an essential ingredient to a master bathroom. Most men prefer showers to baths, so much so that the lack of a shower can be an impediment to selling a house. At the same time, separating the bathtub and shower gives the room a sense of elegance.

In my own master bath I placed the tub by the windows overlooking the garden on one side of the room, and located the shower on the other side of the room next to the lavatory. My shower is 3 × 3 feet, dimensions I feel are minimal for a master shower; 4 × 4 feet would have been better.

As in all baths and showers, we recommend the use of a pressure-balanced antiscald valve for both comfort and safety. It is no longer uncommon to see two or more shower heads in a shower stall. The extra plumbing will amount to a small percentage of the overall cost of your bathroom, so if you have the space for a larger stall, and your budget can accommodate the increased cost of the tile, shower doors, plumbing, and so on, there is no harm in this. There are many times in my household, and I suspect in yours, too, when my wife and I both want to shower at the same time. Such an arrangement would be welcome; the kids would have a great time in it, too.

As you design your bathroom, you might consider a shower stall with an L-shaped entrance, which would need no door. In the home magazines such designs *look* expensive, but when you consider that a merely acceptable shower door will cost hundreds and a first-class shower door around a thousand dollars, a self-sealing shower stall starts to make economic sense.

Two other devices of note are a computerized faucet with a microprocessor-controlled mixing valve with which you program your ideal shower temperature. Another is an adjustable track-mounted "personal shower," some models of which not only look good, but help you rinse your body and clean the shower stall.

●

The typical bathtub measures 60 inches long by 30 inches wide by about 16 inches deep. These dimensions are suitable for cleaning one's self but not for a good long soak. By the time the water is up to the overflow drain, you can barely submerge yourself in hot water. Moreover, the tubs are designed to be installed flush against walls

on three sides, which makes the tub feel closed in and inhospitable.

What one really wants is a vessel long enough to stretch out in, and deep enough in which to become fully submerged. Additionally, one wants space surrounding the tub for soaps, bath gels, reading material, and simply to avoid the feeling of being hemmed in. It would be nice to be able to look out onto a garden or other natural scene, to calm the soul while soothing the body. The Japanese have their traditional *furos*, or soaking tubs, and the Europeans long, deep bathtubs. American manufacturers are beginning to catch on, though; some of them offer "bathing pools," which are whirlpool baths without the pump and jets. These tubs come in lengths from about 5 to 7 feet, in widths from 3 to 6 feet, and in depths from 18 to 32 inches. Just as with a whirlpool, they can be mounted in a built-up pedestal and given the necessary surrounding space and accoutrements.

Very large tubs, like large whirlpools, may require structural modifications to the floor joists to carry the additional load of the tub, water, and occupants. Using all the hot water in a water heater at once shortens the life of the unit, so many large tubs and whirlpools may require a separate water heater to fill the tub.

But, despite the extra work and trouble, an adequately sized, set-in tub with a carefully designed surround will not only be great to use but will become one of the key design features of the room.

The gold-plated tub (yes, real gold) was original to the apartment, and an entirely new bathroom built around it. Vaulted ceilings, marble mosaic floor, and *tromp l'oeil* frescos are intended to create a Pompeiian theme.

(Above) Power bath over-
looks the great space to
capture views of the Ver-
mont countryside. Cur-
tains can be drawn
across the plate-glass
windows for privacy.

(Right) Just in case you
don't buy the Pompeii
recreation, here is the
original — a bath in the
Roman House of Menan-
der (circa 2 B.C.); the fres-
coes have faded a bit, but
the splendor remains.

Whenever "This Old House" has done a master bedroom in the last five years, our homeowners, without exception, have wanted a whirlpool bath. I can hardly blame them; I wanted one, too, and coughed up the money to install a top-of-the-line cast-iron model.

No doubt the swirling water and silver bubbles are soothing, but you have to ask yourself if it's worth the cost. In my own household, my wife rarely uses the whirlpool, although she does take frequent bubble baths. I use the whirlpool perhaps ten times a year. If I had it to do over again, I would install the same unit, in the same manner, *without* the whirlpool apparatus.

After polling Jock Gifford, who has designed many of our "This Old House" projects, "This Old House" producer/director Russ Morash, and several of our "This Old House" homeowners who insisted on a whirlpool, I find, without exception, that no one would install a whirlpool again. We have nothing against them, you must understand, but it seems that they are rarely used.

In making the decision among a tub, a whirlpool, or a bathing pool for your master bath, here are several things to consider:

- Are you a shower person or a bath person? If you rarely take baths, perhaps you'd be better off getting a luxurious, extra-roomy shower and sticking with a standard tub.
- How well does the tub fit you? Climb into a few, probably on the showroom floor at the plumbing supply house (no matter how ridiculous this may make you feel).
- Do you plan to bathe alone or with company? The tubs are sized accordingly. Be aware that the bigger the tub, the longer it takes to fill and the more hot water it will require. If it's likely you will be using the tub alone, you might consider a smaller model.
- Finally, will the soaking tub or whirlpool be used by other members of the household? If so, consider whether it should be installed in a family bathroom. That would reduce intrusions on your privacy in the master suite. (For more information about whirlpools, see the exercise bath chapter.)

Circulation and Focal Points

In most bathroom layouts, the fixtures are placed along the perimeter of the room, leaving the center for circulation. But there is no hard-and-fast rule about this. In a master bathroom "This Old House" did in a house in Arlington, Massachusetts, we employed the opposite arrangement, massing the fixtures in the center and

This master bath with Japanese design elements imparts a sense of spaciousness by separating sinks, toilet, and soaking tub, which gazes out to the meditation garden.

leaving the exterior walls free. The main reason for this was to avoid changing the fenestration, thereby leaving intact the façade of that wonderful old Greek Revival. The floor plan created an L-shaped circulation area, which actually improved one's ability to move about the room, and placed the tub in the quiet part of the L. As we said above, challenging spaces sometimes bring serendipitous design results. And if you are faced with a tortured space in which to place your master bath, try to look for hidden opportunities.

From an aesthetic point of view, it's nice to have a focal point to grab the eye as you enter the room; this can be a well-situated tub, or perhaps a bank of windows to take the eye outdoors. If the toilet is to be in a compartment of its own, you would want to place the compartment out of the way.

Two contrasting uses of stone. Polished granite floor and counter play off the stainless-steel sinks, wall tiles, and glass blocks for a slick, urbane look in a city apartment, whereas a *onsen*-like atmosphere is promoted by the stone wall and smooth wood in the master bath of this suburban home.

Rich colors and detailing set a strong mood.

Materials

As with all baths, the materials you select for the floor, walls, and finishes of your master bath should be handsome, durable, and easy to maintain. Unlike other bathrooms in the house, though, you will probably be willing to spring for more expensive materials here.

But when you are choosing your counter, cabinet, floor, and shower stall materials, keep in mind that it is not merely the quality of the materials but their arrangement and treatment that will give you the overall look you want. In other words, design and detailing are just as, if not more, important than the materials themselves.

When my wife and I were detailing our master bathroom, we, like everyone else, looked into tiling it with marble. Gorgeous and classy, marble also commands a high price, however, both for the tiles and for the labor to install them. We didn't feel the space and the overall architecture of the house warranted the cost, so instead we used a charcoal-glazed tile, and set it off with white grout. The tile setter carefully aligned the grout lines from the shower threshold, across the floor to the dais on which the tub is set. All trim, from the bullnose sanitary base to the door threshold, tub surround, and shower door casing, is done in the same tile and white grout, giving the room a very crisp, expensive look.

Heating and Ventilation

As we said in the chapter on getting the job built, the Cadillac of heat delivery systems is radiant hydronic heat. I myself have lived with every kind of heating system, from blubber fires (in canvas-walled tents while doing research in the Arctic) to forced hot air, hydronic baseboard, and electric. Nothing I have lived with compares to the quality of warmth delivered by radiant hydronic heat. If you have a hot water system in your house, and you are planning to do a significant master bedroom renovation, then explore using radiant floor heat in the bath.

Radiant systems entail running plastic tubing in the floor, which gives off an even, draftless heat. In a bathroom, though, the floor area is often too small to run enough tubing to produce the required BTUs to heat the room. In the master bathroom in the adobe rehab "This Old House" did in Santa Fe, New Mexico, we embedded radiant hot water tubing in the floor, covered by flagstone, and installed more of the tubing in the walls surrounding the whirlpool tub, which we covered with built-up marble paneling.

As superb as radiant floor heating is, it is not for the shallow of pocket. When I renovated my house in 1986, I installed a standard

hydronic baseboard system. Recently, I upgraded the system with a radiant heating zone in the kitchen. I installed the system with my friend and plumber Richard Bilo, so I saved some money on labor, but even with a contractor's discount on the materials, the cost just for the kitchen alone came to $1,500. Granted, the master bathroom is a much smaller area, and it would, therefore, cost less to install radiant, but it is still one of the most expensive options. Still, so comfortable is this heat delivery system that I would probably make room for it in my budget even if I had to sacrifice elsewhere in the project.

As a practical matter, most bathrooms must rely on more conventional delivery means, such as forced hot air ducts, baseboards, or radiators. If you have forced hot air in your house, it

A designer's *tour de force.* This bath is very sleek and powerful with its mixture of granite and marble, glass block, and neon and low-voltage halogen lighting, but I have to wonder how *comfortable* one feels using it.

makes little economic sense not to extend it to serve the new master bath.

If you have a hydronic heating system, and do not want the expense and trouble of upgrading the master bathroom to radiant floor heat, consider, as we suggested above, using the very stylish radiators manufactured by Runtal, DeLonghi, and Acova. Some models do dual duty as towel warmers.

If your heating system is an old one, controlled by a single thermostat, the occasion of renovating your master bath may be a good time to split your heating system into separately controlled zones. When "This Old House" renovated an 1815 Federal-style house in Wayland, Massachusetts, we were faced with a decrepit steam boiler that distributed its heat throughout the house via cast-

iron radiators. Because this was a historic house, we did not want to open up walls to run new piping for separate zones. Instead, our plumbing and heating expert, Richard Trethewey, retrofitted each cast-iron radiator as a separate zone by running a thin, flexible plastic return pipe from each radiator back down to the new hot water boiler. In the master bath, we used a radiant floor system installed by running the tubing underneath the subfloor (while the ceiling of the kitchen just beneath was still open). Incidentally, this is an excellent way to retrofit radiant floor heat in a bath. Usually the cost of opening up a ceiling to install the tubing is far less than attacking the tile walls of the bath itself.

●

As I mentioned before, when I renovated the bathrooms in my current house, an 1836 Colonial Revival, I did not install ventilation fans because of the intrusion on the architecture and the noise. I can now state without qualification this was a mistake. Not only is one obliged to open a window to expel any odors, but humidity from showers, baths, and shaving can build up, too.

A temple to conspicuous consumption, this one has it all: VCR, *three* TVs, two showers, a bathtub, and a whirlpool (beneath the canopy). Through the wooden doors at the back is an "environmental habitat" that provides steam, sauna, sun, wind, or warm rain at the touch of a button.

Rustic materials and construction equal closeness to nature in this Adirondack mountain house. Skylight connects with the outdoors without sacrificing the room's intimacy.

A mechanical ventilation system, vented to the outdoors, is essential. If you object to the noise and the looks of the fan housing, there are units in which the fan itself is located outside the room, such as in the attic or a closet, and connected to the bath via ductwork. You can then cover the ventilation intake with a simple, inconspicuous grille. We recommend putting the fan on a timer so that it will turn off automatically. We also like to avoid a combination fan-and-light switch, as it is annoying to have the noise of the fan when you want only light.

●

The remarks I made at the conclusion of the chapter on the master suite are applicable here, too. Bathrooms in general are design-intensive areas that require careful attention to space planning and detail. Since the master bath is the one bath in the house where you are likely to splurge, and the one bath you claim as your private preserve, get very involved in the design and planning of this room. As in most renovation projects, good results are born of the hard work and involvement of you as the homeowner, as well as the rest of the construction team.

Upon Completion

You should recognize that no matter how carefully you design your bathroom, you are bound to make mistakes. You may catch some of these as you build your project and you can then set them right. Others though, will not become apparent until the job is done and you actually begin to use the baths — the counters should have been two inches higher; you should have installed a single-lever faucet; the tile color for the shower was not quite right. I constantly ponder how I would do my bathrooms differently if I had the chance. I kick myself for not making the master bath larger by eliminating an adjacent closet. Then I could have installed two sinks instead of one. In the guest and children's baths, I simply tiled around existing windows in the middle of the shower wall. Now, I wish I had taken the little extra time to pull those windows out and replace them either with skylights or with clerestory windows high on the exterior wall.

The thing to remember in building and renovation is that you are constructing the prototype. If you and your designer or architect did your work well, you probably considered the design consequences of many alternatives and made what you thought was the best decision at the time. Now, having built the mock-up, you know what you would do if you had it to do over again; but, of course, you probably won't have that chance.

I quite enjoy the process of design, building, and postmortem evaluation, even when the completed projects are not "perfect." If there is anything I have learned from my own endeavors and my involvement with "This Old House" it is that there really is no such thing as "perfection" in building and renovation — just relatively more or less elegant solutions to specific problems. Even more interesting is the fact that the nature of the design problems — and, therefore, the solutions — evolve as we grow older, our habits change, our kids get bigger, and our interests shift.

I begin each project with the expectation that it will be my *last* such project; I finish with a sense of relief. Thereafter, as I have the chance to live in the space I have helped to create, gain more experience working on other projects, and acquire more ideas from colleagues, books, and magazines, I find myself looking forward to renovating the next house, helping to design and build the next bath. I used to be dismayed by this, but it seems to me now that in this way design and renovation mirror the vital process of life itself — which is to say change, evolution, renewal.

Great good luck on your project from all of us at "This Old House."

●

When I grow up and can afford the time and money to build my ultimate house, I'll have a bathroom like this one. Taut, cool, understated design wields the expressive properties of glass, stone, tile, and wood with great confidence. The soaking tub feels like a deep pool at the edge of a mountain abyss.

258

Acknowledgments

260 We wish to thank the many people and organizations who contributed to this book:

Armstrong World Industries; Glenn Berger, Acton Woodworks; Scott Broney and Christy Stadelmaier at American Olean Tile; Martin J. Gill at American Standard, Inc.; Jay Warren Bright, AIA; Ellen Cheever, CKD, National Kitchen and Bath Association; Jack Cronin, Cronin Cabinets; Design Media Resource; Eastern Paralyzed Veterans Association; Gay Fly, ASID, CBD; Ann E. Grasso; Jock Gifford and Jennifer Shakespeare at Design Associates; B. Leslie Hart, *Kitchen and Bath Business;* Helo Sauna and Fitness, Inc.; Richard Howard; Kohler Co.; Dick Metchear; Janine J. Newlin, ASID; and Paul Vogan, Belmont Flooring.

Don Cutler managed the project; Pamela Hartford was our Illustrations Editor; Chris Pullman designed the book. The folks at Design Associates of Cambridge and Nantucket took much time from more profitable ventures to contribute to this book, notably John Murphy, who drew the illustrations, and Chris Dallmus, who did the design on the concept bathrooms and managed the development of the drawings. René Varrin, Paul Lasner of Specialized Housing, Inc., and George Welsh of Living Design were of invaluable aid in preparing the chapter on barrier-free bathrooms.

No list of acknowledgments would be complete without thanking the "This Old House" team, especially Russ Morash, Norm Abram, and Rich Trethewey. I must also thank my wife, Evy Blum, who tolerated the intrusion work on this manuscript made into our family time. As always, when I think about house design, the memory of Marilyn Ruben is very much with me. Mentor, friend, and surrogate mother to my wife for more than twenty years, Marilyn designed all of our own renovation projects, including our current house. Her designs were spare, elegant, and above all, superbly functional. This book is dedicated to the memory of my friend and mentor Will Humphreys, who though it was one of life's wonderful surprises that, through great good luck, one of his most earnest philosophy students became host of "This Old House."

Photograph Credits

Cover:
Design: Stephen A. Magliocco Associates, Architects; Boston, MA
Interiors: Nancy Eddy Design; Dedham, MA
Photography: Richard Howard

ii
Design: Josh Schweitzer; Los Angeles, CA
Photography: Tim Street-Porter

vi
Design: Winthrop Faulkner and Partners; Washington, DC
Photography: Norman McGrath

xii
Design: Kliment Halsband; New York, NY
Photography: John Hall

x
Design: Sakakura Associates; Tokyo, Japan
Photography courtesy Hotel Oh-an

xiii
Design: Charles R. Myer; Cambridge, MA
Photography: John Hall

xiv
Design: Brian Murphy; Los Angeles, CA
Photography: Tim Street-Porter

xi
Photography: Kari Haavisto

viii
Photography: Richard Howard

xvi upper right
Photography: Richard Howard

xvi lower right
Photography: Richard Howard

xvi left
Photography: Richard Howard

xvii
Design: William McDonough; New York, NY
Photography: John Hall

2
Design: Richard Beard; San Francisco, CA
Builder: Steve Plath, Plath & Company
Photography: William Helsel

4
Design: Feldman Hagan; New York, NY
Photography: Andrew Hagan

5 left
Design: Feldman Hagan; New York, NY
Photography: Kari Haavisto

5 right
Design: Ace Architects; Berkeley, CA
Photography: Alan Weintraub

6
Design: Gail and Steve Huberman; Woodbury, NY
Photography: Norman McGrath

8
Design: Brian Murphy; Los Angeles, CA
Photography: Tim Street-Porter

10
Design: Stanley Joseph
Photography: Lynn Karlin

11
Design: Clodagh; New York, NY
Photography: Lizzie Himmel

12
Design: Frank Pennino; Los Angeles, CA
Photography: Norman McGrath

14
Design: Mullman Seidman Architects; New York, NY
Photography: John Hall

89
Design: Charles Marks; Greenwich, CT
Photography: Norman McGrath

91
Photography: Tim Street-Porter

92
Photography: Tim Street-Porter

94
Design: Hariri & Hariri; New York, NY
Photography: John Hall

97
Design: Louis Mackall; Guilford, CT
Photography: Robert Perron

100
Design: Eugenia Voorhees and Hugh Newell Jacobson
Photography: John Hall

102
Design: Frank Gravino; New Haven, CT
Photography: Karen Bussolini

105
Design: Robert Bray + Michael Schaible; New York, NY
Photography: Lizzie Himmel

106
Design: Banks Design; Norwalk, CT
Photography: Robert Perron

108
Photography: Norman McGrath

110
Photography: John Hall

112
Photography: Matthew Walker

113
Design: May Yin Architects; Washington, DC
Photography: Bruce Katz

114
Design: Neil Kelly, Builders Remodelers; Seattle, WA
Photography: Alan Burt

116
Design: Pratt & Larsen; Portland, OR
Photography: Laurie Black

117
Design: The Artful Dodger; Port Townsend, WA
Photography: Laurie Black

118
Design: David Esterich and Associates; New York, NY
Photography courtesy David Esterich

119
Design: Classic Design Associates; Corte Madera, CA
Photography: William Helsel

120
Photography: Dennis Krukowski

122
Photography: Kari Haavisto

123
Design: Al Duarte; Palm Springs, CA
Photography: Laurie Black

124
Design: Duo Dickinson; Madison, CT
Photography: Robert Perron

125
Design: Design Associates; Cambridge & Nantucket, MA
Photography courtesy Jennifer Shakespeare

127
Design: Thadani Hetzel Partnership; Washington, DC
Photography: Anne Gummerson

128
Design: Brian Murphy; Los Angeles, CA
Photography: Tim Street-Porter

130
Design: David Esterich and Associates; New York, NY
Photography courtesy David Esterich

131
Design: Daniel Wolpin, AIA; Minneapolis and St. Paul, MN
Photography: Karen Melvin

133
Design: Eric A. Chase, Architect; Branford, CT
Photography: Karen Bussolini

134
Design: Brian Murphy; Los Angeles, CA
Photography: Tim Street-Porter

135
Design: Matthew Smyth; New York, NY
Photography: Dennis Krukowski

136
Photography: Laurie Black

137
Photography: Laurie Black

138
Design: Maria Garces; New York, NY
Photography: Norman McGrath

139
Design: R. T. Mudge; Lyme, NH
Photography: Robert Perron

140
Design: Malcolm Holzman; New York, NY
Photography: Norman McGrath

142
Design: Johnson-Wanzenberg; New York, NY
Photography: John Hall

144 left
Design: Banks Design; Norwalk, CT
Photography: Robert Perron

144 right
Design: Beverly Markus; Minneapolis, MN
Photography: Karen Melvin

146
Design: Banks Design; Norwalk, CT
Photography: Robert Perron

147
Design: Harry Williams; New York, NY
Photography: Norman McGrath

148 left
Design: Rollin B. Child; Plymouth, MN
Photography: Karen Melvin

148 right
Design: Bruce Bierman; New York, NY
Photography: Kari Haavisto

149 left
Design: May Yin Architects; Washington, DC
Photography: Bruce Katz

149 right
Design: Dou Dickinson; Madison, CT
Photography: Mick Hales

150 top
Design: Centerbrook Architects and
Planners; Essex, CT
Photography: Norman McGrath

150 bottom
Photography: Dennis Krukowski

151 top left
Design: Neil Wright; Ketchum, ID
Photography: Norman McGrath

151 top right
Design: Frank Faulkner
Photography: John Hall

151 bottom
Design: Dianne Love; New York, NY
Photography: John Hall

152
Design: Thadani-Hetzel Partnership;
Washington, DC
Photography: Gordon Beall

154
Design: Louis Mackall; Guilford, CT
Photography courtesy Louis Mackall
Plan courtesy Louis Mackall

159
Photography: Laurie Black

160
Design: Nelson Denny
Photography: Karen Bussolini

162
Photography: Laurie Black

163
Design plans courtesy of Brock
Simini Architects; Washington,
DC

168
Photography courtesy Kohler
Company

173
Photography courtesy Kohler
Company

175
Design: Design Associates;
Cambridge, MA
Photography courtesy Jennifer
Shakespeare

181
Photography courtesy KWC
Company

182; 183
Design: George Welsh, Living
Design; Seattle, WA

Photography courtesy George
Welsh

186
Design: Peter Gisolfi Associates;
Hastings-on-Hudson, NY
Photography: Norman McGrath

188
Design: Suellen DeFrancis;
Scarsdale, NY
Photography courtesy Rikugo
Construction

189
Design: Rus Calder; New York,
NY
Cabinetry: St. Charles, NY
Photography: John Schwartz

191
Design: Abraham Rothenberg
Architect; Westport, CT
Photography: Karen Bussolini

192
Design: Frank Faulkner
Photography: John Hall

193
Design: Signature Kitchen and Bath;
West Hartford, CT
Photography: Karen Bussolini

194
Design: Myron Goldfinger; New
York, NY
Photography: Norman McGrath

195
Design: Alan Buchsbaum; New
York, NY
Photography: Norman McGrath

196
Design: Ivan Chermayeff; New York,
NY
Photography: Norman McGrath

198; 199
Design: Stanley Joseph
Photography: Lynn Karlin

201
Photography courtesy Kohler
Company

202
Design: Stuart Conley; New York,
NY
Photography: Todd Henkels

211
Design: Stephen and Gail
Huberman
Photography: Philip Ennis

212
Design: Peter Gisolfi Associates;
Hastings-on-Hudson, NY
Photography: Norman McGrath

214
Design: Brian Murphy; Los Angeles,
CA
Photography: Tim Street-Porter

216
Design: Peter Gisolfi Associates;
Hastings-on-Hudson, NY
Photography: Robert Perron

217
Design: Stein + Associates; Boston,
MA
Photography: Richard Howard

218
Design: Gus Dudley
Photography: Karen Bussolini

219
Design: Pensis Stolz, Inc.; New York,
NY
Photography: Dennis Krukowski

221
Design: Marilyn Ruben/Steve
Thomas
Photography: Richard Howard

222
Design: John Stedilla; New York, NY
Photography: John Hall

223
Design: Harry Williams; New York,
NY
Photography: Norman McGrath

228
Design: David Easton; New York, NY
Photography: Lizzie Himmel

229
Design: Custom Kitchens; Berkeley,
CA
Photography: Laurie Black

230
Photography: John Hall

231 top
Photography: Dennis Krukowski
Courtesy Sagamore Hills National
Historic Trust

231 bottom
Design: Frank Pennino; Los
Angeles, CA
Photography: Norman McGrath

232
Design: Scavello Design; San
Francisco, CA
Photography: Laurie Black

233
Design: Parish Hadley; New York,
NY
Photography: John Hall

234; 235
Design: Allan Wanzenberg; New
York, NY
Photography: John Hall

236; 237
Design: Harry Williams; New York, NY
Photography: Norman McGrath

238
Design: Greg & Wies Architects;
New Haven, CT
Photography: Robert Perron

239 left
Design: Malcolm Holzman; New
York, NY
Photography: Norman McGrath

239 right
Design: Brian Murphy; Los Angeles,
CA
Photography: Tim Street-Porter

240
Design: Fred Clapper for Spitzer &
Associates; New York, NY
Photography: Norman McGrath

242
Design: Kliment Halsband; New
York, NY
Photography: Norman McGrath

243
Design: Mary Meehan; New York,
NY
Photography: Dennis Krukowski

244; 245
Design: Patrick Naggar; Paris and
New York
Photography: John Hall

246
Design: Margaret Helfand; New
York, NY
Photography: Norman McGrath

248
Design: Patrick Naggar; Paris and
New York
Photography: John Hall

249 top
Design: Sandra Nunnerly; New
York, NY
Photo: Jaime Ardiles Arce

249 bottom
Photography courtesy The
Gallagher Collection

251
Design: Brukoff Design; Sausolito,
CA
Photography: Michael Bry

252 top
Design: Lois Mirviss; New York, NY
Photography: John Hall

252 bottom
Design: Levin-Brown Architects;
Baltimore, MD
Dvorine Associates Interior
Designers; Baltimore, MD
Photography: Anne Gummerson

253
Design: Sam Botero; New York,
NY
Photography: Phillip Ennis

254; 255
Design: Bromley-Caldari; New York,
NY
Photography: Jaime Ardiles Arce

256
Photography courtesy Kohler
Company

257
Design: Beryl Brown; New York, NY
Photography: Lizzie Himmel

259
Design: Charles Gwathmey; New
York, NY
Photography: Norman McGrath

265

Index